MACROMOLECULAR STRUCTURE AND SPECIFICITY: COMPUTER-ASSISTED MODELING AND APPLICATIONS

ANNALS OF THE NEW YORK ACADEMY OF SCIENCES
Volume 439

MACROMOLECULAR STRUCTURE AND SPECIFICITY: COMPUTER-ASSISTED MODELING AND APPLICATIONS

Edited by Babu Venkataraghavan and Richard J. Feldmann

The New York Academy of Sciences
New York, New York
1985

Library of Congress Cataloging in Publication Data

Main entry under title:

Macromolecular structure and specificity.

    (Annals of the New York Academy of Sciences,
ISSN 0077-8923 ; v. 439)
    Papers presented at a conference which was held by
the New York Academy of Sciences on Oct. 12-14, 1983.
    Bibliography: p.
    Includes index.
    1. Macromolecules—Data processing—Congresses.
I. Venkataraghavan, Babu. II. Feldmann, Richard J.
III. New York Academy of Sciences. IV. Series.
Q11.N5   vol. 439   500 s   85-4951
[QD381.9.E4]   [547.7′028′54]

ISBN 0-89766-272-5
ISBN 0-89766-273-3 (pbk.)

CCP
Printed in the United States of America
ISBN 0-89766-272-5 (Cloth)
ISBN 0-89766-273-3 (Paper)
ISSN 0077-8923

ANNALS OF THE NEW YORK ACADEMY OF SCIENCES

Volume 439
March 30, 1985

MACROMOLECULAR STRUCTURE AND SPECIFICITY: COMPUTER-ASSISTED MODELING AND APPLICATIONS[a]

Editors and Conference Organizers
BABU VENKATARAGHAVAN AND RICHARD J. FELDMANN

CONTENTS

Preface. *By* BABU VENKATARAGHAVAN and RICHARD J. FELDMANN vii

Part I. Topology and Architecture of Macromolecules

Deduction of Binding Site Structure from Ligand Binding Data. *By* GORDON M. CRIPPEN 1

Part II. Computer-assisted Macromolecular Structure Generation: Extension of Existing Information

On the Construction of Computer Models of Proteins by the Extension of Crystallographic Structures. *By* RICHARD J. FELDMANN, DAVID H. BING, MICHAEL POTTER, CHARLES MAINHART, BRUCE FURIE, BARBARA C. FURIE, and LYNN H. CAPORALE 12

Protein Structure and Function by Comparative Model Building. *By* JONATHAN GREER 44

Generation of Nucleic Acid Structures and Binding of Molecules to DNA. *By* K. J. MILLER, FREDERICK H. REIN, ERIC R. TAYLOR, and PAUL KOWALCZYK 64

Part III. Molecular Dynamics of Macromolecules

Molecular Dynamics and Minimum Energy Conformations of GnRH and Analogs: A Methodology for Computer-aided Drug Design. *By* R. S. STRUTHERS, J. RIVIER, and A. T. HAGLER 81

Engineering Aspects of Protein Structure. *By* F. R. SALEMME 97

Dynamic Aspects of Protein Structure. *By* MARTIN KARPLUS 107

Part IV. Applications of Macromolecular Specificity to Drug Design

Modeling the Mechanism of Peptide Cleavage by Thermolysin. *By* DAVID G. HANGAUER, PETER GUND, JOSEPH D. ANDOSE, BRUCE L. BUSH, EUGENE M. FLUDER, EUGENE F. MCINTYRE, and GRAHAM M. SMITH 124

[a] The papers in this volume were presented at a conference entitled Macromolecular Structure and Specificity: Computer Assisted Modeling and Applications, which was held by the New York Academy of Sciences on October 12–14, 1983.

Design of Anticancer Drugs Using Modeling Techniques. *By* V. N. Balaji, J. Scott Dixon, D. H. Smith, R. Venkataraghavan, and K. C. Murdock   140

Structure-Activity Studies: A Three-Dimensional Probe of Receptor Specificity. *By* Garland R. Marshall 162

**Part V. Impact of Computer Hardware and Graphics Technology:
Future Perspectives**

Calculations of the Three-Dimensional Structures of Proteins. *By* Harold A. Scheraga 170

A Machine Architecture for Molecular Dynamics: The Systolic Loop. *By* Neil S. Ostlund and Robert A. Whiteside 195

Index of Contributors 209

Papers Presented at the Conference but Not Submitted for Publication:

Part I.

The Relevance of Protein Architecture to the Design of New Proteins. *By* Jane S. Richardson

Conformation and Conformational Change in Proteins. *By* Cyrus Chothia

Part II.

The Use of One- and Two-Dimensional NMR Spectroscopy to Monitor the Consequences of Amino Acid Replacements in Proteins. *By* J. L. Markley, William M. Westler, Warren C. Bogard, Jr., Michael Laskowski, Jr., Xinjie Wu, Gilberto Ortiz-Polo, and Richard J. Feldmann

Part V.

The Future of Molecular Computer Graphics or Blinding Speed Is of Little Use without Illuminating Interaction. *By* Robert Langridge

Financial assistance was received from:
- BRISTOL-MYERS COMPANY
- BURROUGHS WELLCOME CO.
- CIBA-GEIGY CORPORATION
- E. I. du PONT de NEMOURS & COMPANY
  BIOCHEMICALS DEPARTMENT
  CENTRAL RESEARCH AND DEVELOPMENT DEPARTMENT
- LEDERLE LABORATORIES
- LILLY RESEARCH LABORATORIES
- MERCK SHARP & DOHME RESEARCH LABORATORIES
- NATIONAL INSTITUTES OF HEALTH
- SANDOZ, INC./PHARMACEUTICAL DIVISION, RESEARCH AND DEVELOPMENT
- UNITED STATES ARMY MEDICAL RESEARCH AND DEVELOPMENT COMMAND
- UNITED STATES OFFICE OF NAVAL RESEARCH

Preface

BABU VENKATARAGHAVAN

Research Data Processing
Lederle Laboratories
Pearl River, New York 10965

RICHARD J. FELDMANN

Division of Computer Research and Technology
National Institutes of Health
Bethesda, Maryland 20205

Macromolecular Structure and Specificity: Computer-Assisted Modeling and Applications, a conference sponsored by the New York Academy of Sciences, brought together research and applications-oriented scientists to discuss progress, current status, and future directions in this exciting field.

The conference, held on October 12–14, 1983, consisted of five sessions. They covered the topology and architecture of macromolecules, computer-assisted macromolecular structure generation, molecular dynamics, applications of macromolecular specificity to drug design, and the impact of computer hardware and graphics technology on the study of macromolecular interactions. The topics as chosen and arranged proceeded from an overview of basic science of macromolecular architecture to consideration of current methodologies (with attention paid to the merits and demerits of each one) and perspectives on the future. In addition to providing a forum in which experts could discuss results of the latest research, the conference also gave newcomers to the field an opportunity to learn a great deal in a short time.

This volume is divided into five parts corresponding to the five conference sessions.

Topology and Architecture of Macromolecules. This session focused on the structure, classification, and geometric characteristics of proteins and nucleic acids. Emphasis was placed on the influence of structure on specificity and selectivity.

Computer-assisted Macromolecular Structure Generation: Extension of Existing Information. Crystal structure data have been used to understand the effect of structure on function. Computer techniques have recently been used to apply existing data to the generation of three-dimensional structures of proteins for which crystal structures are not available. This protein structure extension technique has enabled researchers to investigate certain problems not amenable to more conventional methods.

Molecular Dynamics of Macromolecules. Interactions of molecules in biological systems induce changes in shape and characteristics, resulting in specific responses. Recent advances in both experimental and computer-assisted techniques have enhanced our understanding of such changes. Particular systems have been studied, and hypotheses that point the way to further research have emerged.

Applications of Macromolecular Specificity to Drug Design. The size, shape, and energetic and electronic aspects of molecules have been shown to play an important role in recognition and interaction of drugs within the biosystem. Rational methodologies employing these concepts have been proposed for the design and development of new drugs. Computer graphics has been used to visualize, validate, and optimize the design process. Although clear-cut answers are not yet possible, computer-assisted techniques can shed light on these relevant factors.

Impact of Computer Hardware and Graphics Technology: Future Perspectives. The demands on computer power and hardware response for real-time interaction have been increasing steadily with the development of new techniques for the study of macromolecular interactions and specificity. Increasingly powerful computer systems are being used to perform complex operations. High-performance color graphics facilities are used quite extensively to visualize and manipulate results.

Advances in our understanding of interactions at the molecular level and the exciting concurrent developments in biotechnology have provided us with unprecedented opportunities for solving problems related to human health. Rational approaches to the design and development of bioactive compounds will be the preferred mode for scientists engaged in research. Extrapolation of existing knowledge into unknown biological domains using computer techniques will be a spur to human endeavor. The feasibility of experimental verification of predictions using recombinant DNA technology will provide added incentive for the development of robust tools. Discussion of the wide-ranging topics generated much interest and enthusiasm among the participants.

We want to take this opportunity to thank all the speakers and participants for their contributions. The outstanding help provided by the staff of the New York Academy of Sciences in organizing the conference is gratefully acknowledged. Finally, we thank the government agencies and private sources for their generous financial support.

Deduction of Binding Site Structure from Ligand Binding Data

GORDON M. CRIPPEN

Department of Chemistry
Texas A&M University
College Station, Texas 77843

INTRODUCTION

There are many sorts of experiments that provide information about the conformation of protein or other macromolecules, but some of the most abundant sources are difficult to interpret. In particular, it is relatively easy to measure the specific binding constants for a large series of compounds binding to the same "receptor" site on a protein molecule, and the data clearly have a bearing on the shape and chemical nature of the binding site. However, it is not clear just what conclusions may be drawn from such data, and how to do it. I will present an extremely simple, yet physically reasonable, framework for describing the site and its interactions with the various ligand molecules, and then show the simplest situations where structural deductions are warranted.

DATA

The task is to learn about the site structure given the binding data for a series of ligands. Thus for every ligand molecule, m, we are given the ΔG_{bind} and the experimental error, or else some equivalent value ($-\log K_I$, $\log I_{\mathrm{so}}$, etc.), which we will denote by E_m. Since experiments are never exact, each E_m really amounts to a range:

$$E_{-m} \leqslant E_m \leqslant E_{+m}.$$

I will accept a calculated binding energy as agreeing with the observed value if it falls within the appropriate range; otherwise it is not in agreement. Note this is a substantially different approach from the usual least squares fitting procedure, where all is well if the majority of the binding energies agree closely, even at the expense of a few outliers. In contrast, the present attitude is that each observed energy is correct up to its stated error and must be accounted for, regardless of the rest of the ligands.

Energies alone do not determine site shape; the rest of the data lies in the molecular structure of the ligands and the relation of the structure to binding energy. Consider the very small and artificial set of ligands in TABLE 1. Suppose someone showed you the table and asked you to find out, by a series of yes-no questions, which compound he has chosen. The average number of questions you must put to him depends on the selection of compounds comprising the list, and is given by the information theory entropy,[2] η:

$$\eta(Y) = -\sum_i p(Y_i)\log_2 p(Y_i) + \sum_{i<j}\sum p(Y_iY_j)\log_2 p(Y_iY_j) - \ldots$$

where $p(Y_i)$ is the probability of a ligand in the list having structural feature i, and $p(Y_iY_j)$ is the probability of a ligand having simultaneously features i and j. This formula has a number of reasonable properties; because $1\log 1 = 0\log 0 = 0$, any features that always occur or never occur contribute zero to η. Also, suppose we had mistakenly subdivided one feature, Y_a, into two, Y_{a1} and Y_{a2}. Then instead of simply a $-p(Y_a)\log p(Y_a)$ term, there would be $-p(Y_{a1})\log p(Y_{a1}) - p(Y_{a2})\log p(Y_{a2}) + p(Y_{a1}Y_{a2})\log p(Y_{a1}Y_{a2})$. Since features Y_{a1} and Y_{a2} always occur together, $p(Y_{a1}Y_{a2}) = p(Y_{a1}) = p(Y_{a2})$, and the third term cancels the second or first one, giving the same result as before. The easiest way to calculate η (structure) for TABLE 1 is to divide the list's substituents into four mutually exclusive categories: 1–H–2–CH$_3$, 1–CH$_3$–2–H, 1–Cl–2–H, 1–Cl–2–CH$_3$. Then only the first sum applies, and $\eta = 2$ bits. In fact, your questioning scheme could be to ask first if the molecule in question is 1–Cl substituted, and if so, ask about 2–CH$_3$ substitution; otherwise ask about 1–CH$_3$ substitution. Then indeed the compound can be guessed in two yes-or-no questions. An important lesson in this is that although the smallest of the ligands in TABLE 1 has 18 atoms, it is not necessary to go into such detail simply to guess the chosen compound. Instead it is to your advantage to divide mentally each ligand into groups of atoms, such that all the compounds can be assembled from as few different types of groups as possible. For example, such a subdivision is given in the table, in terms of naphthyl, 1–methyl, 2–methyl, and 1–chloro groups. Because the naphthyl moiety is common to all, you need waste no questions concerning it. It is necessary to distinguish between methyl groups at the 1 and 2 positions, as the first two compounds are structural isomers. An algorithm for parsimoniously subdividing a set of molecules has already been described.[1]

If one needs to explain the binding of a set of compounds in terms of some model, then the amount of information per compound that can pos-

TABLE 1. Structural Entropy of a Sample Data Set

Compound	Positional composition			
	naphthyl	1-methyl	2-methyl	1-chloro
1-methylnaphthalene	1	1	0	0
2-methylnaphthalene	1	0	1	0
1-chloronaphthalene	1	0	0	1
1-chloro-2-methylnaphthalene	1	0	1	1

sibly be contained in their structures for the purpose of determining the site is just η. The full atomic detail of the molecular structure does not apply, but rather only the structural variety expressed in the data set. We can do no better than to lump together into groups any cluster of atoms recurring in the given list. For the remainder of this paper, I shall assume that this grouping has already been done, and the molecular structures are described in terms of abstract cluster labels: A, B, *etc.*

MODEL

I wish to describe binding in very simple and general terms, yet with some physical justification. Subdivide all space into a finite number of mutually exclusive regions according to the Voronoi polyhedra method.[3] This consists of choosing a number of "site points," $\{si\}$, $i = 1,...,n_s$, positioning them somehow in space, and assigning to the region of each si the set of points closer to it than to all other sj. Depending on how many site points there are and how they are arranged, some of the regions may have infinite volume. Any one region contains its site point, is convex, and has planar boundaries with its neighboring regions. Some regions of the model may correspond to the solvent, others to sterically inaccessible walls of the binding site, and others to energetically attractive pockets of various types. A concave part of the real site, such as a binding cavity, can be represented by the union of two or more adjacent regions (for example, see FIGURE 6, regions 2 and 4). A ligand molecule is idealized as a collection of points in space, each one corresponding to an atomic cluster as described in the previous section. Conformationally flexible molecules will perhaps permit some sorts of motions of the "ligand points" with respect to one another, but I will not explicitly include the internal conformational energy. Each ligand point has a type, denoted by A, B,..., Z, but not all ligand points in one molecule have to have distinct types. Thus structural isomers are distinguished by the relative positions of the ligand points. Whenever a ligand point of type A lies in the region associated with site point $s1$, they are said to be in contact, and the contact makes an additive contribution, $\varepsilon_{A,s1}$, to the total calculated binding energy. More than one ligand point can be in contact with one site point, but because the regions are nonoverlapping and space-filling, one and only one site point is always in contact with a given ligand point. Thus the binding mode can be given in terms of which site point is in contact with each one of the ligand points. Of the many geometrically and conformationally allowed modes, the calculated one is that mode having the lowest calculated binding energy, just as the real drug molecule seeks the state of lowest free energy over all possible translations, rotations and conformations. The calculated binding energy, E_{calc}, is the one corresponding to the lowest energy mode. There may be more than one mode having energy E_{calc}, analogous to the real molecule having some configurational disorder even when bound to the site. We wish to find the positions of the site points and the ε's such that

$$E_{-m} \leqslant E_{calc,m} \leqslant E_{+m}$$

for all ligand molecules, m. If a solution exists at all for a given data set, there are infinitely many solutions produced by adding additional remote and energetically repulsive site points. Therefore, we seek the solution(s) with the minimal number of site points, n_s. Even given that n_s is minimal, there may be many choices of the ε's and the site point coordinates such that the inequalities above are satisfied. We desire the solution with the greatest number of energetically satisfactory binding modes allowed for each molecule. The criteria of minimal n_s and maximal number of modes amounts to solving the problem without specifying too much of the site structure and overdetermining the positions of the ligands in the site. Thus we seek a maximal entropy solution, in both the information theory and thermodynamic senses.

RESULTS

The simplest possible site would be $n_s = 1$, where all space is encompassed by the region of $s1$. This is the extremely simple physical picture of a single region of indefinite size where all groups of a given type contribute equally to the binding regardless of structural isomerism. Such behavior might be expected when the E_m's are determined largely by the partitioning of the ligands between two phases, but it is hardly indicative of specific binding. There are no degrees of freedom necessary to specify this structureless site, so only the ε's need be determined by the set of inequalities:

$$
\begin{aligned}
r_{A,1}\varepsilon_{A,1} + r_{B,1}\varepsilon_{B,1} \cdots &\geqslant E_{-1} \\
-r_{A,1}\varepsilon_{A,1} - r_{B,1}\varepsilon_{B,1} \cdots &\geqslant -E_{+1} \\
r_{A,2}\varepsilon_{A,1} + r_{B,2}\varepsilon_{B,1} \cdots &\geqslant E_{-2}
\end{aligned}
$$

where $r_{A,1}$ is the number of ligand points of type A in molecule 1, *etc.*, and $\varepsilon_{A,1}$ is the interaction parameter between type A and site point 1. Systems of linear inequalities are eminently solvable by linear programming or subgradient optimization.[4] If the system above (expressed in matrix form as $R\varepsilon \geqslant E$) has no solution, then and only then is its dual consistent[4]:

$$
yR = 0; \quad yE > 0; \quad y \geqslant 0.
$$

Thus we may unambiguously determine whether there is a set of ε's satisfying the data, and if so, what their values are. The ease of solution of sets of linear inequalities is the major motivation behind assuming contacts make additive contributions to the binding energy. If the inequalities are consistent, then we have found the simplest conceivable site model to explain the observations.

Suppose now the inequalities above have no solution. Then we must begin to introduce structure into the site. For given site point coordinates, there will be a set of geometrically allowed binding modes for each molecule, m, and for each mode, the interaction energies must satisfy the inequality

$$\Sigma_\ell \varepsilon_{t\ell,s\ell} \geqslant E_{-m}$$

where ℓ refers to each ligand point in the molecule, $t\ell$ is its type, and $s\ell$ is the site point it is in contact with in that particular binding mode. In addition, there must be at least one mode (and perhaps more than one) for which

$$\Sigma_\ell \varepsilon_{t\ell,s\ell} \leqslant E_{+m}$$

must hold. These are the energetically optimal binding mode(s) of ligand m. If there is no choice of optimal mode for each molecule that corresponds to a consistent set of inequalities, then the given site structure can not explain the observations. Although in real applications one is concerned with several ligands, I will restrict these examples to the binding of two molecules only, in order to exhibit the simplest possible situations where site structure is required. Clearly, if accounting for the binding of two molecules out of a large data set requires a certain amount of site structure, then the whole data set will require no less structure. Let R_n denote n-dimensional Euclidean space. Of course, we are really interested only in R_3, but it will be very helpful to examine also the line and the plane.

Suppose in R_1 we have the enantiomeric pair, A–B ($E_1 = -5 \pm 0.5$) and B–A ($E_2 = -8 \pm 0.5$). It is not important what the binding energies are, other than that they are irreconcilably different. Neither does it matter that the $d_{A,B}$ distances in the two molecules are the same or different. It is essential that group types A and B are distinguishable. The two molecules are different in R_1 because they cannot employ the second dimension to rotate and interconvert. Clearly $n_s = 1$ is sufficient only if $E_1 = E_2$. FIGURE 1 shows the three binding modes available to each for $n_s = 2$. If the interaction energies, $\varepsilon_{i,j}$, are chosen as in TABLE 2, then the minimal energy binding modes agree with the observed values. Note that choosing $\varepsilon_{A,1} > 0$ would also give a solution in complete agreement with the given energies, but molecule A–B would then be restricted to only a single binding mode.

FIGURE 1. All possible binding modes of the enantiomers A–B and B–A in R_1, when $n_s = 2$. The circled modes are the energetically optimal ones corresponding to the interactions in TABLE 2.

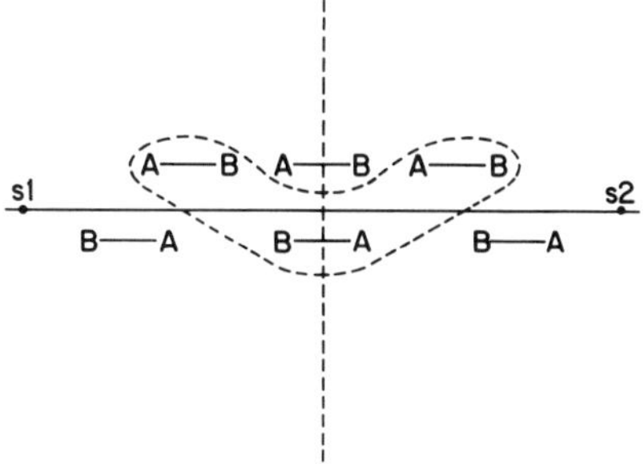

TABLE 2. Optimal Interaction Energy Parameters for Ligands A–B ($E = -5$) and B–A ($E = -8$) in R_1[a]

	Site point	
Ligand point	$s1$	$s2$
A	0	-3
B	-5	-2

[a] Binding modes are shown in FIG. 1.

The choice of ε's in TABLE 2 is maximally noncommittal. Requiring the optimal binding modes shown in FIGURE 1 and requiring an exact match of the given E_1 and E_2 leaves only one degree of freedom in the choice of the four ε's.

In R_2, A–B and B–A cannot be distinguished because they can interconvert. However, the R_1-diastereomers, A–B–A ($E = -5$) and A–A–B ($E = -8$), are distinguishable in any dimensional space, and they require $n_s = 2$. Their only important structural features are that A and B are different and both molecules are linear. In particular, the intramolecular distances are immaterial and need not be similar between the two ligands. Indeed the connectivity suggested by writing dashes between the letters does not imply chemical bonds. The essential difference is that in A–B–A, one must move in opposite directions to reach the A's starting from the B, whereas in A–A–B, the movement is in the same direction. (In R_1, being to the left or right corresponds to left and right handedness in R_3.) The site is constructed by choosing the coordinates of the two site points, $s1$ and $s2$, drawing the line between them, and erecting the perpendicular bisecting plane (in R_3) of that line as the boundary between the two regions. Because we are only interested in how the ligand points fall across the boundary and not in their coordinates, the coordinates of $s1$ and $s2$ do not matter, and there are no degrees of freedom employed in the site structure. FIGURE 2 shows only the energetically optimal binding modes of the two ligands when their interactions are given by TABLE 3. The essential stratagem employed is that A–A–B can push its B group into the $s2$ region and leave both its A's in the $s1$ region. On the other hand, the linear molecule A–B–A must have an A in the same region that its B group occupies because the boundary is planar. This holds as long as $\varepsilon_{A,s1} < \varepsilon_{A,s2}$, and $\varepsilon_{B,s1}$ and $\varepsilon_{B,s2}$ are determined by fitting E_1 and E_2. Therefore there are two degrees of energetic freedom remaining in TABLE 3.

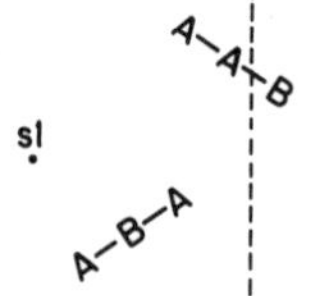

FIGURE 2. The energetically optimal binding modes of the diastereomers A–B–A and A–A–B in R_3, when $n_s = 2$. The interactions are given in TABLE 3.

TABLE 3. Optimal Interaction Energy Parameters for Ligands A–B–A ($E = -5$) and A–A–B ($E = -8$) in R_3[a]

Ligand point	Site point	
	$s1$	$s2$
A	-2	$+1$
B	-1	-4

[a] Binding modes are shown in FIG. 2.

The next most complicated stereochemical problem is that of enantiomers in R_2. There the equivalent of handedness is clockwise versus counterclockwise orientation of three different ligand points, A, B, and C, as shown in FIGURE 3. Suppose the first, counterclockwise molecule has $E_1 = -5$, and the second, clockwise one has $E_2 = -8$. In the plane they can rotate and translate, but because they cannot flip without going out of the plane, they are distinguishable. Otherwise their conformations are unimportant. No site with $n_s = 2$ is sufficient, since any mode preferred by one of the ligands would be available and also energetically optimal for the other. Therefore $n_s = 3$ is the minimal site required, but as before the details of the placement of $s1$, $s2$, and $s3$ are immaterial. TABLE 4 gives a satisfactory set of ε's to account for the given energies. The first ligand could be translated and rotated to put two of its three groups into the regions the second ligand prefers, but it can never simultaneously bring all three to the correct regions. There are a number of different optimal solutions corresponding to relabeling the regions and the ligand points, but having two modes for the first ligand is apparently the highest entropy possible. The four zero entries in TABLE 4 could be increased, but the other entries are fixed by fitting the data for the three modes shown in FIGURE 3, while keeping all other modes at higher energy. Enantiomeric pairs in R_3 can be handled in an exactly analogous fashion: each consists of four distinct groups, ABCD, arranged to be nonplanar, and choosing $n_s = 4$ creates four regions meeting at a point. The interaction energies are chosen so that the stronger binding ligand has each of its four ligand points in a different region for optimal energy, and any other matching of group and region is energetically penalized.

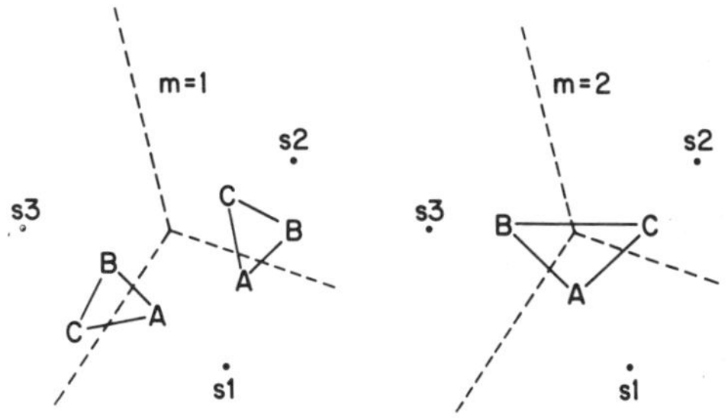

FIGURE 3. The energetically optimal binding modes of the ABC enantiomers in R_2, when $n_s = 3$; the interactions are given in TABLE 4. Ligand 1 on the left has ABC in counterclockwise order, and ligand 2 on the right has clockwise orientation.

TABLE 4. Optimal Interaction Energy Parameters for ABC Enantiomers in R_2 Having $E_1 = -5$ and $E_2 = -8^a$

	Site point		
Ligand point	$s1$	$s2$	$s3$
A	-2.5	-0	-0
B	0	$+0.25$	-2.75
C	0	-2.75	$+0.25$

^{a} Binding modes are shown in Fig. 3.

We have seen how the differential binding of R_1-diastereomers can be explained in terms of a two-site point model. Very similarly R_2-diastereomers in R_3 can have different binding energies only if $n_s \geq 3$. Observe in FIGURE 4 that ligand $m = 1$ can be thought of as two fused ABC triplets, one with clockwise orientation and the other with counterclockwise. However, ligand 2 consists of two counterclockwise triplets. Their planarity is the only important conformational feature otherwise. Of course, such compounds are usually referred to as *cis-trans* isomers, but their relation to the A–B–A and A–A–B pair considered above makes it easy to predict that $n_s = 3$ will be sufficient. FIGURE 4 views the respective optimal binding modes from above the plane defined by the coordinates of $s1$, $s2$, and $s3$. Otherwise their coordinates are unimportant, as long as they are not collinear. If $E_1 = -5$ and $E_2 = -8$, then the interaction energies for a solution are given in TABLE 5. Ligand 2 has a unique optimal mode, but ligand 1 has contacts between (A,B,C,A) and four respective sets of site points: (2,2,1,3), (1,2,1,3), (3,2,1,2), and (3,2,1,1). Thus there are at least five equalities determining the entries of TABLE 5. This model would not necessarily work if the dihedral angle between two of the planes was greater than 180°, but that can never happen with Voronoi polyhedra constructed from only three points.

All the examples considered so far had no detailed conformational features, as are so often encountered in real drug binding studies. Suppose in R_3 we have two A–B molecules, a short one with $d_{A,B} = 10$ Å and $E_1 = -6$, and a long one with $d_{A,B} = 20$ Å and $E_2 = -8$. Clearly $n_s = 2$ is insufficient (all modes available to one can also be achieved by the other, and

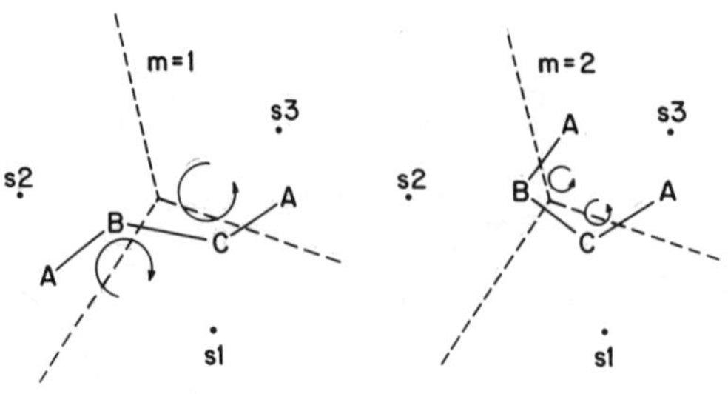

FIGURE 4. The energetically optimal binding modes of *cis–trans* isomers in R_3, when $n_s = 3$ and the interactions are given in TABLE 5.

TABLE 5. Optimal Interaction Energy Parameters for *Cis-Trans* Isomers in R_3 Having $E_1 = -5$ and $E_2 = -8^a$

	Site point		
Ligand point	$s1$	$s2$	$s3$
A	0.7	0.7	-2.3
B	0	-1.7	1.3
C	-1.7	0	1.3

^{a} Binding modes are shown in FIG. 4.

hence their optimal binding energies will always be the same), but $n_s = 3$ works when $s1$, $s2$, and $s3$ are collinear and

$$10 \text{ Å} < \frac{d_{s2,s3} + d_{s1,s2}}{2} < 20 \text{ Å}.$$

Then, as shown in FIGURE 5, the long molecule can span the unfavorable region 2 whereas the short one cannot. Note that at last, the site construction specifies two degrees of freedom: collinearity of the site points and the inequality above. The interaction energies are given in TABLE 6.

If there is less ligand point type differentiation to help explain the binding, a heavier load falls on the site structure. Thus if the B ligand point of the previous example is replaced by A, then the three-point site of FIGURE 5 will no longer suffice. In order to account for the better binding of the long isomer, $\varepsilon_{A,s1} + \varepsilon_{A,s3} = -8$, and if, say, $\varepsilon_{A,s1} \leqslant \varepsilon_{A,s3}$, then the short isomer can bind with both its A's in region 1 and achieve as low an energy as the long one can. For similar reasons, $n_s = 3$ and 4 also fail, but if $n_s = 5$ a solution can be found in R_2 by constructing region 1 with maximum diameter less than the short bond length but having region 5 just close enough to 1 for the long isomer to span across region 4 (see FIGURE 6). Letting $\varepsilon_{A,s1} = -6$ and $\varepsilon_{A,s5} = -2$ and all the others be zero, the long molecule has one most favorable mode available to it, and the short one has three. Raising any of $\varepsilon_{A,s2}$, $\varepsilon_{A,s3}$, or $\varepsilon_{A,s4}$ above zero serves only to reduce the number of energetically optimal modes available. Curiously, reversing the given binding energies, so that the short molecule binds better, yields an easier problem, requiring only $n_s = 4$ in R_2. The trick is to construct a region 1 large enough to accommodate both A's of the short molecule but small enough

FIGURE 5. The energetically optimal binding modes of long and short A–B molecules in R_3 with $n_s = 3$ and the interactions are given in TABLE 6. The short molecule has four energetically optimal modes.

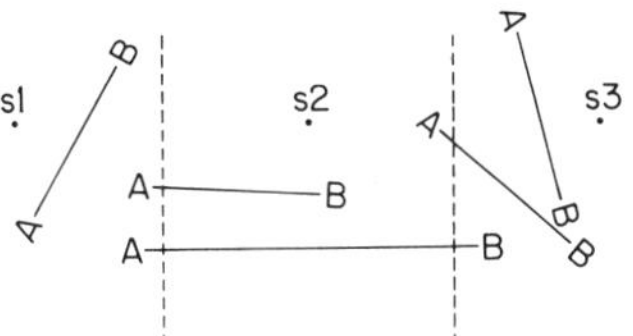

TABLE 6. Optimal Interaction Energy Parameters for Short and Long A–B Molecules Having $E_1 = -6$ and $E_2 = -8^a$

Ligand point	Site point		
	$s1$	$s2$	$s3$
A	-4	-2	-2
B	-2	-2	-4

^{a} Binding modes are shown in FIG. 5.

to force the long molecule to put one of its ends into a different region. Then $\varepsilon_{A,s1} = -4$ and $\varepsilon_{A,s2} = \varepsilon_{A,s3} = \varepsilon_{A,s4} = -2$ gives a maximum entropy solution.

CONCLUSIONS

I have presented a simple class of models of drug binding that are adequate to account for nonspecific binding, conformational and steric effects, and even stereospecific binding. It is closely enough related to the generally successful standard distance geometry approach to analyzing drug binding that one can assume it will be capable of dealing with many ligands, either chemically similar or diverse, that are actually found in the literature. The Voronoi polyhedra approach produces site configurations that are much simpler in many ways (all space is accounted for, the boundaries are planar, and the regions are convex), so that one might hope for a more rigorous and thorough analysis of possible models to explain a given data set. The binding modes produced in this work restrict the molecular coordinates less than does the equivalent distance geometry site model in terms of small spherical site points. If efficient algorithms can be found for examining successively more complicated site geometries, we will have a means of unambiguously finding maximum entropy binding site models.

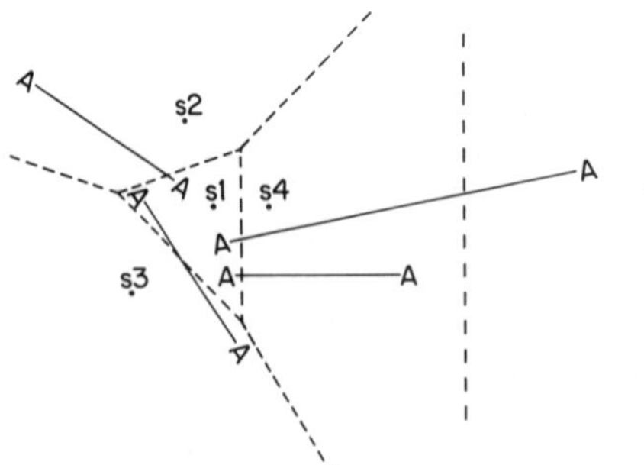

FIGURE 6. The energetically optimal binding modes of long and short A–A molecules in R_2 with $n_S = 5$. $E_{short} = -5$, $E_{long} = -8$, $\varepsilon_{A,s1} = -5$, $\varepsilon_{A,s5} = -3$, and the other ε's $= 0$. The short molecule has three distinct but equally favorable lowest energy modes.

REFERENCES

1. CRIPPEN, G. M. 1981. Distance Geometry and Conformational Calculations. Wiley. New York.
2. COX, R. T. 1961. The Algebra of Probable Inference. Johns Hopkins Press. Baltimore.
3. VORONOI, G. F. 1908. J. Reine Angew. Math. **134:** 198–287.
4. SANDI, C. 1979. *In* Combinatorial Optimization. N. Christofides, A. Mingozzi, P. Toth & C. Sandi, Eds. Wiley. New York.

On the Construction of Computer Models of Proteins by the Extension of Crystallographic Structures

RICHARD J. FELDMANN,[a] DAVID H. BING,[b]
MICHAEL POTTER,[c] CHARLES MAINHART,[c]
BRUCE FURIE,[d] BARBARA C. FURIE,[d]
AND LYNN H. CAPORALE[e]

[a] *Division of Computer Research and Technology
National Institutes of Health
Bethesda, Maryland 20205*

[b] *Cambridge Research Laboratory
Cambridge, Massachusetts 02139
and
Beth Israel Hospital
Harvard Medical School
Cambridge, Massachusetts 02139*

[c] *Laboratory of Cell Biology
National Cancer Institute
National Institutes of Health
Bethesda, Maryland 20205*

[d] *Department of Medicine
Tufts New England Medical Center
Boston, Massachusetts 02111*

[e] *Department of Biochemistry
Georgetown University Medical School
Washington, D.C. 20007*

INTRODUCTION

One of the central beliefs of molecular biology is that the sequence of amino acids determines the structure of a protein.[2-4] Yet our scientific culture has still to discover many of the rules that specify and govern this relationship.

The results of X-ray crystallographic determination of specific proteins have led to an understanding of the taxonomy of protein architectures.[4] The protein sequence file[5] which had been growing rather slowly because each amino acid sequence had to be determined by chemical techniques is now growing exponentially because amino acid sequences can be derived from nucleic acid sequences.[6] Protein extension attempts to shorten the delay between the generation of the protein sequence file and the construction of a protein model based on the crystallographic data. Such models of three-dimensional structure may be less precise but do facilitate the design of biochemical experiments that require such a model.

This exercise also offers insight and some understanding of the relationship between amino acid sequence and the three-dimensional structure of a protein. Statistical analysis of the secondary structure features of proteins for which there is a crystal solution to three-dimensional structure has led at best to only partial success in predicting the three-dimensional structure of proteins in general.[7-12] Admittedly the statistical sample (*i.e.*, conformational properties of dipeptides and tripeptides in crystallographically resolved structures) is rather small with respect to the state space of all possible proteins.*ƒ* Because energy minimization of protein structure still takes hundreds if not thousands of hours of computational time on even modern laboratory computers, it has been difficult to build up an empirical base of knowledge that relates sequence to three-dimensional structure. Analysis of the protein sequence file shows that in families of proteins the amino acid sequences are highly conserved. The key to the protein extension problem lay in realizing that the folding pattern of a protein family is more conserved than the amino acid sequence. Conflict arose because sequence matching techniques for the most part ignore the three-dimensional placement of amino acids.

We used a simplified model of proteins (*i.e.*, one sphere per amino acid) to explore the location of amino acid differences in various proteins in relation to the crystal structure. The results indicated that, for the most part, the differences lay on the surface of the protein. The idea soon emerged that the function of the surface of a protein is to present a unique structure which might give an evolutionary advantage to the protein and that the notion of preserving the three-dimensional structure is not necessarily compatible with the statistical notions used in secondary and tertiary protein prediction problems since approximately half of the amino acids in a small globular protein are exposed to the solvent. With the pressure off absolute amino acid sequence equivalence we came to realize that in this situation only the general properties of a protein have to be maintained (*i.e.*, charge balance or charge versus hydrophobicity balance). We noticed that the structural core of a protein (*i.e.*, beta barrels and alpha helix-beta sheet-alpha helix composites) is constituted by amino acids which are part of the entire sequence. The highest homology seemed to exist in the core regions and the noncore portion of a protein was usually a collection of loops between helices and beta sheet strands. In these loops the homology was weak and the length of loops was quite variable. The graphical modeling of loops posed some problems which will be explained in the body of the paper. In considering structure function relationships of the various families of proteins, however, it appeared to us

---

ƒ The combinatorics of possible protein sequences is 20! × *n*!, so that if *n* is 150, which is typical for a compact single domain globular protein, the state space is 20! × 150!. This is a big number.

that either in natural evolution or in the basic engineering requirements for the function of the protein, loop modification was one of the easiest ways of making major modifications to a protein. It was only later that Fletterick[13,14] showed that the exon-exon interface in the DNA maps onto the loops and from an engineering viewpoint this was exactly what one would want to be true. In the meantime we built a software system that permitted the construction of any member of any protein family for which a crystal structure for one member of that family existed. Construction of a model of a protein with no changes in amino acid length was done automatically. A whole family of length-conserved proteins (for example ovomucoids or cytochrome c's) could be done in an afternoon. Proteins with significant length changes required two or three days of work where the graphical modeling and human error were the rate-limiting steps.

In this paper the technical details of our methods are explained, but it is important to remember that the success of the method (for which as yet no conventional scientific evidence exists) depends on viewing the proteins as whole entities in the context of natural evolution. The wholeness of a protein is a careful balance between folding, function, amino acid sequence, architecture and component parts. Protein extension is a technical means for producing models of families of proteins, and a potential vehicle for designing proteins. It combines high-tech hardware and flexible software to explore the relationship between primary and tertiary structure.[15]

MULTIPLE SEQUENCE ALIGNMENT

A collection of amino acid sequences form a family if there is sufficient homology in the sequence to make a multiple alignment of the amino acids. Until recently automatic alignment of even pairs of amino acid sequences by a computer program has been very slow. The time required to make a pairwise match increases as the cube of the number of amino acids being matched increases. Lipman and Wylbur[16] have produced a rapid program which matches pair of sequences in $n \times \log n$ time but there is not yet a program which will align multiple sequences efficiently. A function MULTIPLE-SEQUENCE-INPUT in the program EDISEQ (from EDItor of SEQuences) enables the manual alignment of a family of sequences. FIGURES 1–3 show the results of this function. The commands within the function permit the shifting of rows to create a global alignment of the sequences. The row shifts represent insertions and deletions with respect to the amino acid sequence in the crystal structure used for the molecule. Other commands permit the exchange of columns to help in the homology ordering of the proteins. The least homologous sequences are moved to the left.

In each of FIGURES 1–3 the sequence runs from the N-terminal at the top to the C-terminal at the bottom. The amino acid sequence of the protein for which there is a crystal solution is the first sequence at the right. The secondary features of the crystal structure are shown to the right of the se-

```
                3         2         1
     4321098765432109876543210987654321
 1   AHGGMGGGGGGGGGGVGVGGGGGGGGGGGGGGGG    1
 2   DDLLELLLLLLLLLLLLLLLLLLLLLLLLLLLLLL    2
 3   FASSLNSSSSSSSSSSSSSSSSSSSSSSSSSSS     3   Alpha Helix
 4   DEDDSDDDDDDDDDDDEEDDDDDDDDDDDDDDDD     4   Alpha Helix
 5   ALEDDQGGGGGGGGGGAAGAGGGGGGGGGGGGG     5   Alpha Helix
 6   VVEEQEEEEEEEEEEEEEEEEEEEEEEEEEEEEEE    6   Alpha Helix
 7   LLWWEWWWWWWWWWWWWWWWWWWWWWWWWWWWWW     7   Alpha Helix
 8   KKKHWQQQQQQQQQQEQQQQQQQQQHQQQQQQQQ     8   Alpha Helix
 9   CCKHKQQLLLLALLQLLLLLLLSLLLLILLLLLL     9   Alpha Helix
10   WWVVHVVVVVVVVVVVVVVVVVVVVVVVVVVVVVV   10   Alpha Helix
11   GGVLVLLLLLLLLLLLLLLLLLLLLLGLLLLLL    11   Alpha Helix
12   PGDGLTTKKNNNNNNNKHHNNNNKNNNNNNNNHN   12   Alpha Helix
13   VVIIDMIVVIAAAVVTVVIVVVIVVVVVIIVVVV   13   Alpha Helix
14   EEWWINWWWWWWWWWWWWWWWWWWWWWWWWWWWW    14   Alpha Helix
15   AAGAWGGGGGGGGGGGAAAGGGGGGGGGGGGGGG   15   Alpha Helix
16   DDKKTKKKKKKKKKKKKKKKKKKKKKKKKKKKKK   16   Alpha Helix
17   YFVVKVVVVVVVVVVVVVVVVVVVVVVVVVVVVVV   17   Alpha Helix
18   TEEEVEEEEEEEEEEEEEEEEEEEEEEEEEEEEEE   18   Alpha Helix
19   TGPPESAGTTAAAAAAAAAAAAAAAATATAAAAA   19
20   MTDDSDDDDDDDDDDDDDDDDDDDDDDDDDDDDDD   20   Alpha Helix
21   GGLLKLILIEIVVVIIVLVLLVLLVILLLLIIII   21   Alpha Helix
22   GGPSLAAPTGPAAAAPASAAAGAAAPAVAAPPPP   22   Alpha Helix
23   LESAPGGGGGGGGGGGGGGGGGGGGGGGGGGSSG   23   Alpha Helix
24   VVHHEHHHHHHHHHHHHHHHHHHHHHHHHHHHHHH   24   Alpha Helix
25   LLGGHGGGGGGGGGGGGGGGGGGGGGGGGGGGGGG   25   Alpha Helix
26   TTQQGHHQQKQQQQQEQQQQQQQQQQQQQQQQQQ   26   Alpha Helix
27   RREEHAEEDDEEEEEFDEDDEEDEEEEEEEEEEE   27   Alpha Helix
28   LLVVEVVVVVVVVVVVIIIVVVVIVVVVVVVVVV   28   Alpha Helix
29   FFIIVLLLLLLLLLLLLLLLLLLLLLLLLLLLLLL   29   Alpha Helix
30   KKIIMMIIIIIIIIVIIIIIIIIIIIIIIIIIII   30   Alpha Helix
31   EQRRIRRRRRRRRRRRRRRRRRRRRRRRRRRSRR   31   Alpha Helix
32   HHMLRLLLLLLLLLLLLLLLLLLLLLLLLLLLLLL   32   Alpha Helix
33   PPFFLFFFFFFFFFFFFFFFFFFFFFFFFFFFFFF   33   Alpha Helix
34   EEQQLKHKKKKTTKTTKKKKKHTTTKKKKKKKKK   34   Alpha Helix
35   TTNVQSDTTGGGGGGGSGGGTGAAGGSGNGDGGG   35   Alpha Helix
36   QQHHEHHHHHHHHHHHHHHHHHHHHHHHHHHHHHH   36   Alpha Helix
37   KKPPHPPPPPPPPPPPPPPPPPPPPPPPPPPPPPP   37   Alpha Helix
38   LLEEPEEEEEEEEEEEEEEEEEEEEEEEEEEEEEE   38   Alpha Helix
39   FFTTETTTTTTTTTTTTTTTTTTTTTTTTTTTTTT   39   Alpha Helix
40   PPQQTMLLLLLLLLLLLLLLLLLLLLLLLLLLLLLL  40   Alpha Helix
41   KKDEQDDEEEEEEEEEEEEEEEEEEEEEDEEEEE   41   Alpha Helix
42   FFRRERRKKKKKKKKKKKKKKKKKKKKKKKKKKK   42   Alpha Helix
43   AVFFRFFFFFFFFFFFFFFFFFFFFFFFFFFFFFF   43
     --------------------------------------------------------
     4321098765432109876543210987654321   Sequence number
                3         2         1
```

FIGURE 1. Thirty-four myoglobin sequences, numbered from right to left. The line numbers on left and right (1 to 43) match because there are no changes in the length of the myoglobin, yet it is a highly conserved structure.

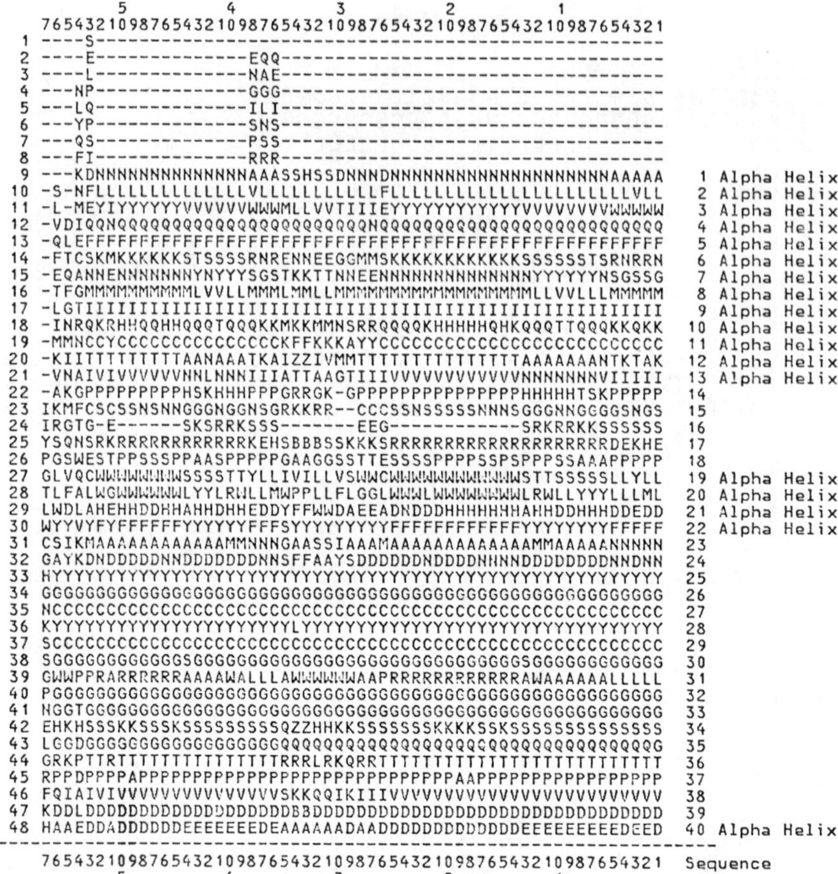

FIGURE 2. Fifty-seven phospholipase a sequences, numbered from right to left. The line numbers on the right are for the member of the family which has a crystal solution. The numbers on the left differ according to the occurrence of insertions or deletions (see text).

quences. In FIGURE 1 (myoglobin) there are no insertions or deletions so that the number of the crystal structure sequence on the right is the same as the line number on the left. The line number is used to specify the manipulating actions in the MULTIPLE-SEQUENCE-INPUT function. In FIGURES 2 and 3 (snake neurotoxins vs. bovine phospholipase a2) the crystal structure numbers on the right and the line numbers on the left differ because there are extensive insertions and deletions. The ordering of the sequences from right to left are a measure of homology to the crystal structure amino acid sequence. The left-most sequences are generally very difficult to align.

The portion of the myoglobin alignment shown in FIGURE 1 is typical of the whole alignment in that there is very clear homology at many places in the sequence across almost the whole family. The N-terminal portion of the phospholipase a2 sequences shown in FIGURE 2 also shows a clear pattern of homology at many places in the sequence across almost the entire family. Notice at crystal structure position 16 (line number 24) there are some sequences with deletions. These deletions occur in a loop region between two helices. Lines 1–8 for sequences 36–38, 53 and 54 are *N*-terminal insertions since there is no equivalent amino acid sequence in the crystal structure. These amino acids could be integrated into the structure by extending the *N*-terminal alpha helix. The *C*-terminal portion of the phospholipase a2 family shown in FIGURE 3 shows many more gaps. There are extensive insertions and deletions across the family. The cysteine residues (represented in the one letter code by C) have been used for alignment since the chemical and crystal structures are known to have identical disulfide bonding of cysteine pairs. Line

```
          5         4         3         2         1
765432109876543210987654321098765432109876543210987654321
60 PYYGYYYYYYYYYYYYYYYYYYYYYYYYYYYYYYYYYYYYYYYYYYYYYYYYYYYYYYY  52 Alpha Helix
61 NGGNASGEGDNGDAGGGSDTKRGGGGGGGGGANDDNEEGGGGDSASDGGGGGRKTRK  53 Alpha Helix
62 VKKIEDDKEEEEEREEEDEQQDKKKKKKKDDEEEEEKKEEEEEDRDEEEEEEDQQDQ  54 Alpha Helix
63 MAMYAAAAAAAAAAAAAAAAAAAVLLAAMAAAAAAAAAAAAAAAAAAAAAAAAAAAAA  55 Alpha Helix
64 STG-----------------------------------------------------
65 ANTNGEE--EEEET-----KKK-------EEGEEEG----EEEET-------KKKKK  56 Alpha Helix
66 GCYLKKKGEKKKKKEEEEGEKN-------KKKKKKKGGEEKKKKKEGEEEEEENKENK  57 Alpha Helix
67 ENDALIKKKIIIISKKKKKLLL-------KKLIIIIKKKKIIIISKKKKKKKLLLLL  58 Alpha Helix
68 SPTASSHMLSSSSYMMQKKSDDTAATTGDHHSSSSSMMLLSSSSYKKQQMMMDDSDD  59 Alpha Helix
69 KKKKAGKGGRGGGSGGGGGSSSGKKDNTKKKAGGRGGGGGGGGGSGGGGGGGSSSSS  60 Alpha Helix
70 HTWCCCCCCCCCCCCCCCCCCCCYCCCCCCCCCCCCCCCCCCCCCCCCCCCCCCCCCC  61 Alpha Helix
71 GVTCKRNWWWWWWTYYYSFRKK---------KWWWWWWWWWWWRTSFYYYYYKKRKK  62 Alpha Helix
73 TYYSV---------------LLL----------V-----------------LLLLL  64 Alpha Helix
74 DTNPL---------------VVV----------L----------------VVVVV  65
75 TYYNS---------------DDDDBB------S----------------DDDDD  66
76 ASERE-P----PP--P---NNNBTTNNDSNNE--------PPP--------PNNNNN  67
77 SEIKPPKPPPPYYPPKPPPPPFPKKPFTPPPPPPPPPPPPPYYYPPPPPPPPPKPPPPP  68
78 REQTNYTYYYYIIYKWKKKYYYKYYKKKKKKNYYYYYYYYYIIIYYKKKKKKWYYYYY  69
79 LNNYNFSFLFFKKWLTMMMTTTLWWTTWMTTNFFFFLFLLKKKFWMMLMLLTTTTTT  70
80 SGGVDKQTTKKTTTTTLLSSENEDBBVVTISSDKKKKTTTTTTTKTSSLLTTLNNEEN  71
81 CEGYTTYLLTTYYLMYMAASNSVIISSSLQQTTTTTLLLLYYYTLAAMMMYSNSSN  72
82 NIITYYSYYYYTTYYTYYYYYYYYYYYYYYYYYYYYYYYYYYTTTYYYYYYYYTYYYYY  73
83 NVDCSSYKKSSYYSNYDDDKSSTRRTTNSSSSSSSSSKKKKYYYSSDDDDNNYSSKSS  74 Beta Sheet
84 NCCNYYKYYYYDEWYEYYYFYYYYYYYYYYYYYYYYYYYYYYDEEYWYYYYYYEYYFYY  75 Beta Sheet
85 DGDAEDLKEEESSQYSYYYSSSTSSSSEKKKEEEEEKKEESSSDQYYYYYYSSSSSS  76 Beta Sheet
86 FGEP----------------------------------------------------
87 YDDACCTCCCCCCCCCCCCCCCCCELLEEIFLLCCCCCCCCCCCCCCCCCCCCCCCCCC  77 Beta Sheet
88 KDPFNTKSSSSQQIGTGGGSSSEKKEEQHTTNSSSSSSSSSQQQTIGGGGGGTSSSSS  78 Beta Sheet
89 NPQGEKRQQQQGGETDSEEGNNBSSNNHNKKESQQQQQQQGGGKEEESSTTDNNNNN  79
90 SCKIGGTGGGGTTKZTNNNTNTGGGGGGGRRGQGGGGGGGGTTTGKNNNNQQTTTTTN  80 Beta Sheet
91 AGKKQKIKKTTLLTSSGGGEEEAYYEEGNTTQGTTTKKKKLLLKTGGGGSSSEEEEE  31 Beta Sheet
92 DTETLLILLLLTTPPPPPPVIIIIIIIIIIIIILTLLLLLLLLTTTLPPPPPPPPIIVII  82 Beta Sheet
93 TQLVTT-TTTTS-TY-YYYTTTVTTIVDVIITLTTTTTTTSS-TTYYYYTT-TTTTT  83 Beta Sheet
94 II-------------------------T-----------------------------
95 SCCCCCCCCCCCCCCCCCCCCCCCCCCCCCCCCCCCCCCCCCCCCCCCCCCCCCCCCCCC  84 Beta Sheet
96 SEEDNKYSSKKGKDDDRRRSSNGGGGGDGYYNKKKKSSSSGGKKDRRRRNDDNSSNS  85 Beta Sheet
97 YCCCDEGGGNGADSBENNHNDSSGKKGGEDGGDGGNGGGGGAADESNNNNDDESSDSS  86
98 FDDDDGAGGGGAGKTKVIIKEKBGGDDDKAADGGGDGGGGANGGKIIVVKKKKKEKKE  87
99 VKRR--A--------------------------A----------------------
100 GAVDNNGNNN----K-KKKNNN-------AGN-NNNNNNN---N-KKKK---NNNNN  88
101 KAACDNGSNNNNGTTTKKKNNNBTTDDPNGGDNNNNSSNHNNGNTKKKKTTTNNNNN  89
102 MAAQEETKKANNKGGGKKKAAAXWWPPQATTENAASKKKKNKKEGKKKKGGGAAAAA  90 Alpha Helix
103 YIITCCCCCCCCCCCCCCCCCCCCC-CCCCCCCCCCCCCCCCCCCCCCCCCCCCCC  91 Alpha Helix
104 F---------A--------------------------------------------
105 NCCCKA-GEAAAAQZQNLLEEEXZZGGKK--KAAAAGGAEAAAAQLLNNQQQEEEEE  92 Alpha Helix
106 LFFDAARAAAAAARRGRRRAAAXZZTTKKRRAAAAAAAAAAAAAAARRRRRRRGAAAAA  93 Alpha Helix
107 IRAAFFIAAAASSFYFKFFFFFXZZQQEKVIFAAASAAAASSSFFFFKKYYFFFFFF  94 Alpha Helix
108 NDNYIVVVVVVVVVVVVIIIIIIIILVVVIVVVVVVVVVVVVVVVVVVVVVVIIIII  95 Alpha Helix
109 TNNHCCCCCCCCCCCCCCCCCCCCCCCCCCCCCCCCCCCCCCCCCCCCCCCCCCCCCC  96 Alpha Helix
-----------------------------------------------------------
   765432109876543210987654321098765432109876543210987654321   Sequence
          5         4         3         2         1
```

FIGURE 3. Central portion of the alignment of the phospholipase sequences (see FIGURE 2).

86 in FIGURE 3 is probably an artifact of the poor alignment of the left-most sequences. Line 104 in FIGURE 3 is necessary because sequence 47 is bounded on either side by aligned cysteines at positions 91 and 96.

The representation of the secondary features (alpha helix, beta structure) of the crystal structure sequence after alignment are effectively mapped onto the whole family of sequences. In the simple cases such as myoglobin this mapping means that the three-dimensional crystal structure can be used as a prototype for building a model of any of the members of the family. The multiple alignment tool is very effective in bringing order out of the initial chaos of a family of sequences but the recognition of patterns by the operator is a key and important aspect of achieving a reasonable alignment.

SINGLE SEQUENCE ALIGNMENT

The MULTIPLE-SEQUENCE-INPUT function provides a way of aligning amino acid sequence in a whole family of proteins. Sometimes the family may consist of only two members. The sequence of the protein for which a three-dimensional structure has been determined is the "source" or "master" sequence and the sequence of the protein for which a model is to be constructed is the "target" or "slave" sequence. In other cases the alignment of an individual target sequence extracted from the global alignment needs refinement with respect to the source sequence. The function EDIT-SUBSTITUTION-SEQUENCE in the program EDISEQ provides the means for aligning the target sequence against the source sequence. This function is an editor consisting of a number of commands. The editing proceeds by first giving the command which types out a portion of the sequence alignment; then one or more commands are given to change the alignment. The user goes through this type/alter cycle until the alignment is better.

In FIGURE 4 is the alignment of the *N*-terminal portion of minke whale myoglobin with the crystal structure of sperm whale myoglobin. The source sequence is to the left and the target sequence is to the right. The secondary structure features are presented to the left of the crystal structure sequence. A column of letters (P for polar, A for nonpolar and N for neutral) to the left and right of the two sequences assigns a value as to whether an amino acid is external or internal with regard to the surface of the protein. The column of stars ($\times \times \times$) and plusses (+ + and +) shows the match between the two sequences. Identity of sequence is indicated by three stars ($\times \times \times$). The table in FIGURE 5 shows the pattern of evaluation. For example an ASP which is matched to a GLU is evaluated as (+ +) and an ASP or a GLU which is matched to a LYS or ARG is evaluated as (+). On the other hand a LYS or ARG matched to any member of the class (PHE, PRO, MET, VAL, LEU, or ILE) is evaluated as (− −). This evaluation scheme follows the class assignment given by Dickerson and Geis.[17] The data presented in FIGURE 4 illustrate the close homology exhibited by the minke and sperm whale proteins.

The availability of protein material as well as the ease with which a pro-

	1	P	VAL	VAL	P	1	✳✳✳
	2	P	LEU	LEU	P	2	✳✳✳
Alpha Helix	3	N	SER	SER	N	3	✳✳✳
Alpha Helix	4	A	GLU	ASP	A	4	++
Alpha Helix	5	N	GLY	ALA	N	5	++
Alpha Helix	6	A	GLU	GLU	A	6	✳✳✳
Alpha Helix	7	N	TRP	TRP	N	7	✳✳✳
Alpha Helix	8	N	GLN	HIS	N	8	+
Alpha Helix	9	P	LEU	LEU	P	9	✳✳✳
Alpha Helix	10	P	VAL	VAL	P	10	✳✳✳
Alpha Helix	11	P	LEU	LEU	P	11	✳✳✳
Alpha Helix	12	N	HIS	ASN	N	12	+
Alpha Helix	13	P	VAL	ILE	P	13	++
Alpha Helix	14	N	TRP	TRP	N	14	✳✳✳
Alpha Helix	15	N	ALA	ALA	N	15	✳✳✳
Alpha Helix	16	A	LYS	LYS	A	16	✳✳✳
Alpha Helix	17	P	VAL	VAL	P	17	✳✳✳
Alpha Helix	18	A	GLU	GLU	A	18	✳✳✳
	19	N	ALA	ALA	N	19	✳✳✳
Alpha Helix	20	A	ASP	ASP	A	20	✳✳✳
Alpha Helix	21	P	VAL	VAL	P	21	✳✳✳
Alpha Helix	22	N	ALA	ALA	N	22	✳✳✳
Alpha Helix	23	N	GLY	GLY	N	23	✳✳✳
Alpha Helix	24	N	HIS	HIS	N	24	✳✳✳
Alpha Helix	25	N	GLY	GLY	N	25	✳✳✳
Alpha Helix	26	N	GLN	GLN	N	26	✳✳✳
Alpha Helix	27	A	ASP	ASP	A	27	✳✳✳
Alpha Helix	28	P	ILE	ILE	P	28	✳✳✳
Alpha Helix	29	P	LEU	LEU	P	29	✳✳✳
Alpha Helix	30	P	ILE	ILE	P	30	✳✳✳
Alpha Helix	31	A	ARG	ARG	A	31	✳✳✳
Alpha Helix	32	P	LEU	LEU	P	32	✳✳✳
Alpha Helix	33	P	PHE	PHE	P	33	✳✳✳
Alpha Helix	34	A	LYS	LYS	A	34	✳✳✳
Alpha Helix	35	N	SER	GLY	N	35	+
Alpha Helix	36	N	HIS	HIS	N	36	✳✳✳
Alpha Helix	37	P	PRO	PRO	P	37	✳✳✳
Alpha Helix	38	A	GLU	GLU	A	38	✳✳✳
Alpha Helix	39	N	THR	THR	N	39	✳✳✳
Alpha Helix	40	P	LEU	LEU	P	40	✳✳✳
Alpha Helix	41	A	GLU	GLU	A	41	✳✳✳
Alpha Helix	42	A	LYS	LYS	A	42	✳✳✳
	43	P	PHE	PHE	P	43	✳✳✳
	44	A	ASP	ASP	A	44	✳✳✳
	45	A	ARG	LYS	A	45	++
	46	P	PHE	PHE	P	46	✳✳✳
	47	A	LYS	LYS	A	47	✳✳✳
	48	N	HIS	HIS	N	48	✳✳✳
	49	P	LEU	LEU	P	49	✳✳✳
	50	A	LYS	LYS	A	50	✳✳✳

FIGURE 4. The master is on the left and target is on the right. (✳ ✳ ✳) means total homology. (+ +) and (+) indicate less homology. (See text for details.)

	ASP GLU	TYR	ASN GLN THR SER CYS	HIS	TRP GLY ALA	LYS ARG	PHE PRO MET VAL LEU ILE
ASP GLU	++	+			–	+	––
TYR	+	++	+	+			–
ASN GLN THR SER CYS		+	++	+	+		–
HIS		+	+	++		+	–
TRP GLY ALA	–		+		++	–	++
LYS ARG	+			+	–	++	––
PHE PRO MET VAL LEU ILE	––	–	–	–	++	––	++

FIGURE 5. A matrix indicating how amino acid substitutions are related to each other in the protein extension method. (+ +) is close, (+) is partially close, (–) is small relationship, and (– –) is nonidentity.

tein can be crystallized and the structure resolved dictate which protein sequence will be the master structure. As the protein families are quite conservative with respect to function, the match between sequences and hence the evaluation using the scheme described above is often quite good.

The alignment of a snake toxin with bovine phospholipase a2 in FIGURE 6 shows a number of new features of protein sequence alignment. The pattern of stars ($\times \times \times \times \times$) at the right side of the diagram indicates that the matching disulfide linkages are preserved. The structure presented in FIGURES 17, 20 and 21 uses dots ($\cdots$) to indicate a matching which does not preserve the disulfide linkages. In FIGURE 6 the observed homology is somewhat weaker than that in FIGURE 4. Insertions are used to bring different portions of the target snake toxin sequence into alignment with the phospholipase sequence. The beta strand region between 74 and 85 has a reasonably strong match in sequence homology, as does the match for the region 90 to 109. The loop region 65 to 73 between an alpha helix and a beta strand in FIGURE 7 has a rather poor homology with the crystal member in the pair. The deletion eliminates part of the alpha helix. The actions required to shift the deletion towards the *C*-terminal are presented in FIGURE 7. It can be seen that the structure of the alpha helix is preserved, forcing the deletion out onto the loop region where the three-dimenstional structure is probably more varied.

The process of searching for optimal homology was simplified by making a command that slides the target sequence across the source sequence. In FIGURE 8 the sliding operation and evaluation is performed for the preliminary match of factor B protein to chymotrypsin for a section of sequence which is structured as a beta sheet in chymotrypsin. The original alignment has the maximum homology based on the procedure outlined above and il-

Structure							
Alpha Helix	60	N	SER	GLY	N	60	+
Alpha Helix	61	N	CYS	CYS	N	61	***
Alpha Helix	62	A	LYS	TRP	N	62	−
Alpha Helix	63	P	VAL					
Alpha Helix	64	P	LEU					
	65	P	VAL					
	66	A	ASP					
	67	N	ASN	PRO	P	67	−
	68	P	PRO	TYR	N	68	−
	69	N	TYR	ILE	P	69	−
	70	N	THR	LYS	A	70	
	71	N	ASN	THR	N	71	++
	72	N	ASN	TYR	N	72	+
	73	N	TYR	THR	N	73	+
Beta Sheet	74	N	SER	TYR	N	74	+
Beta Sheet	75	N	TYR	GLU	A	75	+
Beta Sheet	76	N	SER	SER	N	76	***
Beta Sheet	77	N	CYS	CYS	N	77	***
Beta Sheet	78	N	SER	GLN	N	78	++
	79	N	ASN	GLY	N	79	+
Beta Sheet	80	N	ASN	THR	N	80	++
Beta Sheet	81	A	GLU	LEU	P	81	− −
Beta Sheet	82	P	ILE	THR	N	82	−
Beta Sheet	83	N	THR	SER	N	83	++
Beta Sheet	84	N	CYS	CYS	N	84	***
Beta Sheet	85	N	SER	GLY	N	85	+
	86	N	SER	ALA	N	86	+
	87	A	GLU	ASN	N	87	
	88	N	ASN					
	89	N	ASN	ASN	N	89	***
Alpha Helix	90	N	ALA	LYS	A	90	−
Alpha Helix	91	N	CYS	CYS	N	91	***
Alpha Helix	92	A	GLU	ALA	N	92	−
Alpha Helix	93	N	ALA	ALA	N	93	***
Alpha Helix	94	P	PHE	SER	N	94	−
Alpha Helix	95	P	ILE	VAL	P	95	++
Alpha Helix	96	N	CYS	CYS	N	96	***
Alpha Helix	97	N	ASN	ASP	A	97	
Alpha Helix	98	N	CYS	CYS	N	98	***
Alpha Helix	99	A	ASP	ASP	A	99	***
Alpha Helix	100	A	ARG	ARG	A	100	***
Alpha Helix	101	N	ASN	VAL	P	101	−
Alpha Helix	102	N	ALA	ALA	N	102	***
Alpha Helix	103	N	ALA	ALA	N	103	***
Alpha Helix	104	P	ILE	ASN	N	104	−
Alpha Helix	105	N	CYS	CYS	N	105	***
Alpha Helix	106	P	PHE	PHE	P	106	***
Alpha Helix	107	N	SER	ALA	N	107	+
Alpha Helix	108	A	LYS	ARG	A	108	++
	109	P	VAL	ALA	N	109	++

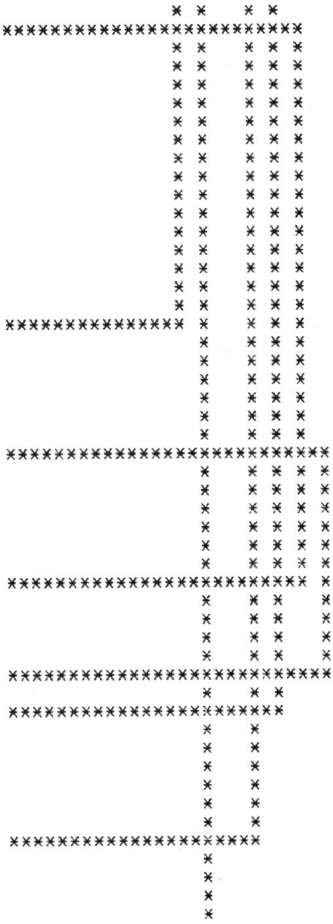

FIGURE 6. The central portion of the alignment of bovine phospholipase with snake toxin.

lustrated in FIGURE 5. This sequence sliding tool is especially useful in establishing piece by piece the global homology between two proteins which have overall low homology (*i.e.*, less than thirty percent homology). One simply tries to find an optimal match between the secondary structural features of the crystal structure and the target protein. The implicit assumption of our method is that the folding pattern of the protein is highly conserved and that insertions and deletions occur in the loop regions between secondary structure features.

GRAPHICAL MODELING

The transformation of the source crystal structure into a model of the target structure involves changing the amino acid side chains to match the sequence

```
--T
Print sequence range     [1-1000] =60,73
Alpha Helix    60   N   SER   GLY   N   60   +              * *     * *
Alpha Helix    61   N   CYS   CYS   N   61   ***    ***************************
Alpha Helix    62   A   LYS   TRP   N   62   -              * *     * * *
Alpha Helix    63   P   VAL                                 * *     * * *
Alpha Helix    64   P   LEU                                 * *     * * *
               65   P   VAL                                 * *     * * *
               66   A   ASP                                 * *     * * *
               67   N   ASN   PRO   P   67   -              * *     * * *
               68   P   PRO   TYR   N   68   -              * *     * * *
               69   N   TYR   ILE   P   69   -              * *     * * *
               70   N   THR   LYS   A   70                  * *     * * *
               71   N   ASN   THR   N   71   ++             * *     * * *
               72   N   ASN   TYR   N   72   +              * *     * * *
               73   N   TYR   THR   N   73   +              * *     * * *
Match value =    4
Match percent = 13.33
Number of modeling actions required = 4
--S
Slave range to be shifted =67,70
Direction and magnitude of shift    (+N or -N or 0 for shiftup) =-4
--T
Print sequence range     [1-1000] =60,73
Alpha Helix    60   N   SER   GLY   N   60   +              * *     * *
Alpha Helix    61   N   CYS   CYS   N   61   ***    **************************
Alpha Helix    62   A   LYS   TRP   N   62   -              * *     * * *
Alpha Helix    63   P   VAL   PRO   P   63   ++             * *     * * *
Alpha Helix    64   P   LEU   TYR   N   64   -              * *     * * *
               65   P   VAL   ILE   P   65   ++             * *     * * *
               66   A   ASP   LYS   A   66   +              * *     * * *
               67   N   ASN                                 * *     * * *
               68   P   PRO                                 * *     * * *
               69   N   TYR                                 * *     * * *
               70   N   THR                                 * *     * * *
               71   N   ASN   THR   N   71   ++             * *     * * *
               72   N   ASN   TYR   N   72   +              * *     * * *
               73   N   TYR   THR   N   73   +              * *     * * *
Match value =   11
Match percent = 36.67
Number of modeling actions required = 4
--
```

FIGURE 7. Actions required to move a deletion within the sequence of the target. This action results in preservation of the alpha helical structure (see text for detail).

of the target. Simple rules for this side chain substitution were developed in the following way. The substituted side chain is put in the shadow of the original side chain. If the original side chain is longer than the substituted side chain then a shortening will generally be easy. If the substituted side chain is considerably longer than the original side chain (for example if a LYS is put in place of an ALA) then the longer side chain uses whatever shadow is available and for the rest of its length is put into an alternating chain conformation which is the preferred conformation for ethylene carbons. Since the conformations of all side chains will eventually be subject to change by the energy minimization procedures, at this point in the construction of a plausible model it is only important to avoid as much steric overlap as possible between various side chains. Conceptually, the graphic modeling should follow the path of change that occurred during natural evolution of the protein, as in a family of proteins whose overall architecture is preserved as a prerequisite for maintenance of function.

In natural evolution, making coordinated changes by random mutation is likely to be a low-probability event and the resulting changes in protein

```
Offset =-1
Beta Sheet      101  N  ASN   GLU   A   100
Beta Sheet      102  A  ASP   GLY   N   101   -
Beta Sheet      103  P  ILE   ASP   A   102   --
Beta Sheet      104  N  THR   ILE   P   103   -
Beta Sheet      105  P  LEU   ALA   N   104   ++
Beta Sheet      106  P  LEU   LEU   P   105   ***
Beta Sheet      107  A  LYS   LEU   P   106   --
Beta Sheet      108  P  LEU   GLU   A   107   --
Beta Sheet      109  N  SER   LEU   P   108   -
Beta Sheet      110  N  THR   GLU   A   109
Match value = -4
--------------------------------------------------
Offset = 0
Beta Sheet      101  N  ASN   GLY   N   101   +
Beta Sheet      102  A  ASP   ASP   A   102   ***
Beta Sheet      103  P  ILE   ILE   P   103   ***
Beta Sheet      104  N  THR   ALA   N   104   +
Beta Sheet      105  P  LEU   LEU   P   105   ***
Beta Sheet      106  P  LEU   LEU   P   106   ***
Beta Sheet      107  A  LYS   GLU   A   107   +
Beta Sheet      108  P  LEU   LEU   P   108   ***
Beta Sheet      109  N  SER   GLU   A   109
Beta Sheet      110  N  THR   ASN   N   110   ++
Match value = 20
--------------------------------------------------
Offset = 1
Beta Sheet      101  N  ASN   ASP   A   102
Beta Sheet      102  A  ASP   ILE   P   103   --
Beta Sheet      103  P  ILE   ALA   N   104   ++
Beta Sheet      104  N  THR   LEU   P   105   -
Beta Sheet      105  P  LEU   LEU   P   106   ***
Beta Sheet      106  P  LEU   GLU   A   107   --
Beta Sheet      107  A  LYS   LEU   P   108   --
Beta Sheet      108  P  LEU   GLU   A   109   --
Beta Sheet      109  N  SER   ASN   N   110   ++
Beta Sheet      110  N  THR   SER   N   111   ++
Match value =   0
--------------------------------------------------
Offset = 2
Beta Sheet      101  N  ASN   ILE   P   103   -
Beta Sheet      102  A  ASP   ALA   N   104   -
Beta Sheet      103  P  ILE   LEU   P   105   ++
Beta Sheet      104  N  THR   LEU   P   106   -
Beta Sheet      105  P  LEU   GLU   A   107   --
Beta Sheet      106  P  LEU   LEU   P   108   ***
Beta Sheet      107  A  LYS   GLU   A   109   +
Beta Sheet      108  P  LEU   ASN   N   110   -
Beta Sheet      109  N  SER   SER   N   111   ***
Beta Sheet      110  N  THR   VAL   P   112   -
Match value =   2
```

FIGURE 8. Sequence sliding to achieve maximum homology using a beta strand from chymotrypsin and factor B protein. The ($\times$) and ($+$) symbols count as $+1$ and the ($-$) symbol counts as -1. The match value is sum of the counts.

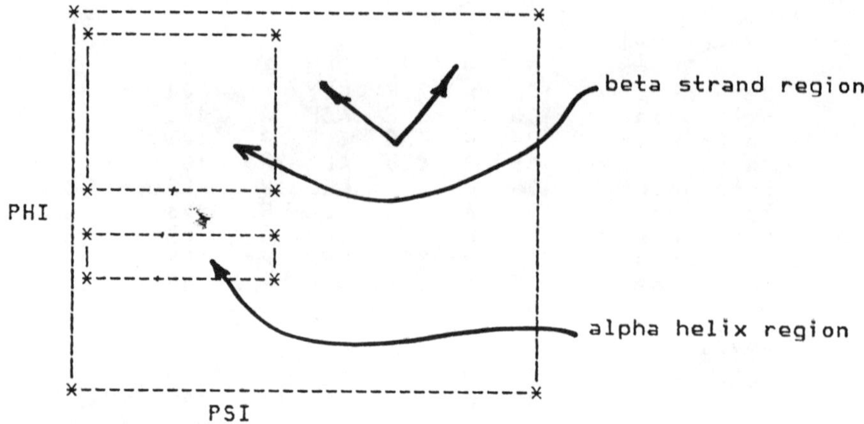

FIGURE 9. Ramaschandran diagram showing reorthogonalization of variables.

structure probably have some selective advantage. In the graphic modeling, we are attempting to proceed rationally to make coordinated changes which will emphasize the similarities rather than the differences. Side chains which face into the center of a molecule are harder to change than side chains which point out into the solvent. If a short side chain facing into the center of a molecule is to be substituted by a long side chain then some concomitant shortening of a spatially adjacent (but not necessarily sequence adjacent) side chain must be accomplished. If an inward facing long side chain is replaced by a shorter one, then the packing of the protein will be looser. Making coordinated changes by random mutations in natural evolution is likely to be a low probability event. However there would be reason to think that in designing new proteins such coordinated changes could be made by a rational decision-making process. In doing so it might be possible to change significantly the characteristics of a protein.

After side chain substitutions are made a test is performed to see if there

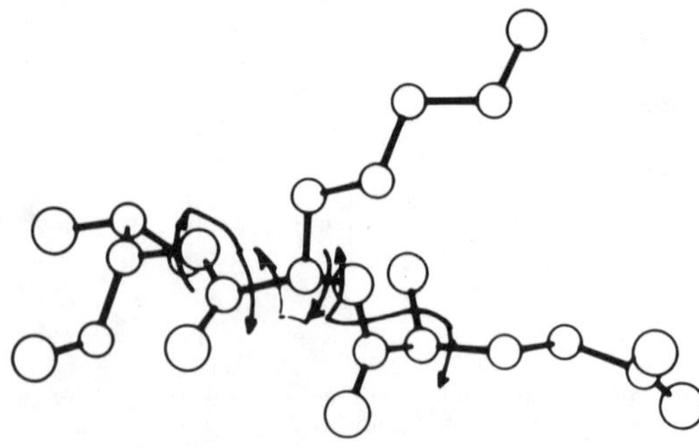

FIGURE 10. Bond mobility for a typical amino acid.

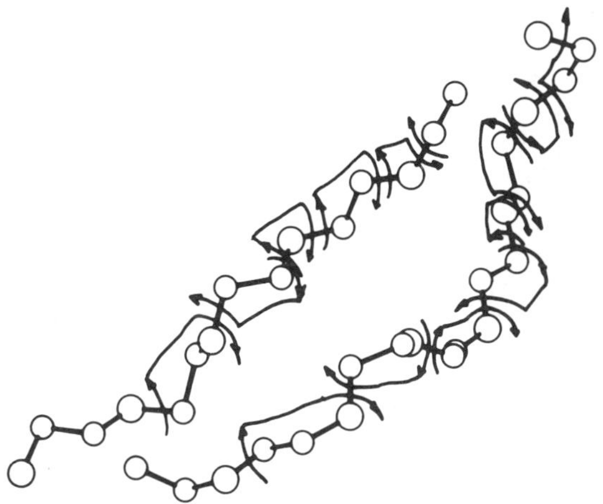

FIGURE 11. Bond mobility for a typical deletion modeling.

are any van der Waal contacts. Whenever these contacts occur, the conformation of the offending side chain is changed to eliminate the contact, as a preliminary form of energy minimization.

Whenever insertions or deletions of sequences of amino acids have to be made into the peptide backbone of the crystal structure there have to be changes in the local conformation of the backbone. The simplest graphical method for making these changes is to associate one modeling operation with each rotatable bond in the region of the peptide backbone being modeled. FIGURE 9 is a generalized Ramaschandran diagram where phi and psi are the conventional names associated with the N-Ca and Ca-C bonds of the peptide backbone.

Initially we tried modeling peptide backbones by varying the phi-psi bonds for four amino acids on either side of the deletion or insertion and found that it was extraordinarily difficult to close up deletions or to open up insertions. The phi and psi variables are ill conditioned for producing subtle changes in peptide backbone conformation. We were just about to give up when one of us (RJF) remembered a comment by David Barry (then at the Computer Systems Laboratory of the Washington University School of Medicine). Barry had tried a difference scheme; in his program when the phi bond is rotated positively the psi bond is rotated negatively by the same amount. To obtain two variables, the same opposite rotation technique is applied to the psi bond of one amino acid and the phi bond of the next *C*-terminal amino acid. This produced the reorthogonalization of the variables. In FIGURE 9 the phi variable is along the *x*-axis and the psi variable is along the *y*-axis while the phi-psi variable is at a − 45 degree angle and the psi-phi variable is at a + 45 degree angle. Both variables in each set are orthogonal to each other.

Modeling of peptide backbones at insertions or deletions using this scheme

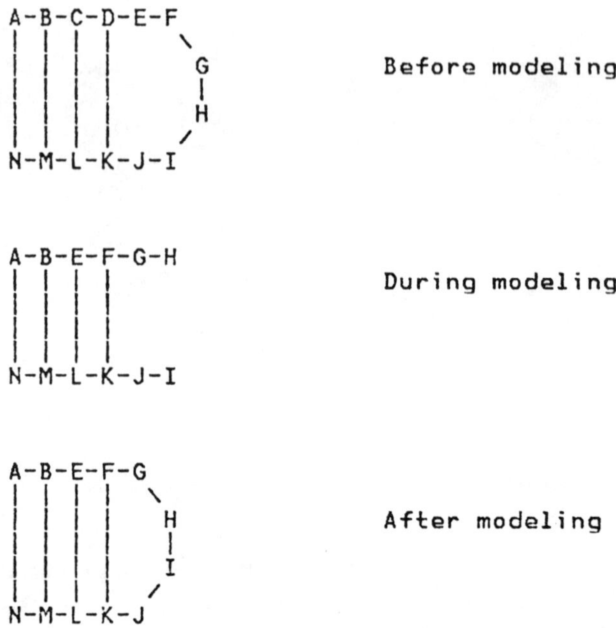

FIGURE 12. Deletion scheme, showing deletion of residues C and D.

was easy. The phi-psi variable changed the curvature of the backbone and the psi-phi variable acted like a crank handle which described a circle of about 1.3 Å radius. The variable assignment for a typical amino acid is shown in FIGURE 10 where the phi-psi difference is shown for the central amino acid and the psi-phi differences are shown for its left and right neighbors. The variable assignment for a typical deletion is shown in FIGURE 11 where the difference variables reach up and down the amino acid backbone on either side of the deletion.

LOOP DELETIONS

Differences in length between the source crystal structure sequence and the target sequence produce insertions and/or deletions that are handled by the graphical methods just described. Each type of change, however, has its special problems and features. The matching of sequences based on homology can produce deletions in the middle of the secondary features. Beta strands, for example, are normally terminated by a loop at either end (the same is true of helices).

In order to close up a deletion in the middle of a beta strand it would be necessary to pull on the beta strand and a portion of the loop structure in such a way as to shorten both the beta strand and the loop. (This is the

Beta Sheet	1	A	ASP	GLU	A	1	++
Beta Sheet	2	P	ILE	ILE	P	2	✳✳✳
Beta Sheet	3	P	VAL	VAL	P	3	✳✳✳
Beta Sheet	4	P	MET	LEU	P	4	++
Beta Sheet	5	N	THR	THR	N	5	✳✳✳
Beta Sheet	6	N	GLN	GLN	N	6	✳✳✳
Beta Sheet	7	N	SER	SER	N	7	✳✳✳
Beta Sheet	8	P	PRO	PRO	P	8	✳✳✳
Beta Sheet	9	N	SER	ALA	N	9	+
Beta Sheet	10	N	SER	ILE	P	10	−
Beta Sheet	11	P	LEU	THR	N	11	−
Beta Sheet	12	N	SER	ALA	N	12	+
	13	P	VAL	ALA	N	13	++
	14	N	SER	SER	N	14	✳✳✳
	15	N	ALA	LEU	P	15	++
	16	N	GLY	GLY	N	16	✳✳✳
Beta Sheet	17	A	GLU	GLN	N	17	
Beta Sheet	18	A	ARG	LYS	A	18	++
Beta Sheet	19	P	VAL	VAL	P	19	✳✳✳
Beta Sheet	20	N	THR	THR	N	20	✳✳✳
Beta Sheet	21	P	MET	ILE	P	21	++
Beta Sheet	22	N	SER	THR	N	22	++
Beta Sheet	23	N	CYS	CYS	N	23	✳✳✳
Beta Sheet	24	A	LYS	SER	N	24	
Beta Sheet	25	N	SER	ALA	N	25	+
Beta Sheet	26	N	SER	SER	N	26	✳✳✳
	27	N	GLN	SER	N	27	++
	28	N	SER	SER	N	28	✳✳✳
	29	P	LEU	VAL	P	29	++
	30	P	LEU	SER	N	30	−
	31	N	ASN	SER	N	31	++
	32	N	SER				
	33	N	GLY				
	34	N	ASN				
	35	N	GLN				
	36	A	LYS				
	37	N	ASN				
	38	P	PHE				
Beta Sheet	39	P	LEU	LEU	P	39	✳✳✳
Beta Sheet	40	N	ALA	HIS	N	40	
Beta Sheet	41	N	TRP	TRP	N	41	✳✳✳
Beta Sheet	42	N	TYR	TYR	N	42	✳✳✳
Beta Sheet	43	N	GLN	GLN	N	43	✳✳✳
Beta Sheet	44	N	GLN	GLN	N	44	✳✳✳
Beta Sheet	45	A	LYS	LYS	A	45	✳✳✳
Beta Sheet	46	P	PRO	SER	N	46	−
Beta Sheet	47	N	GLY	GLY	N	47	✳✳✳
Beta Sheet	48	N	GLN	THR	N	48	++
Beta Sheet	49	P	PRO	SER	N	49	−
Beta Sheet	50	P	PRO	PRO	P	50	✳✳✳

FIGURE 13. The *N*-terminal alignment of McPC–603 immunoglobulin light chain variable domain with the equivalent domain of J539.

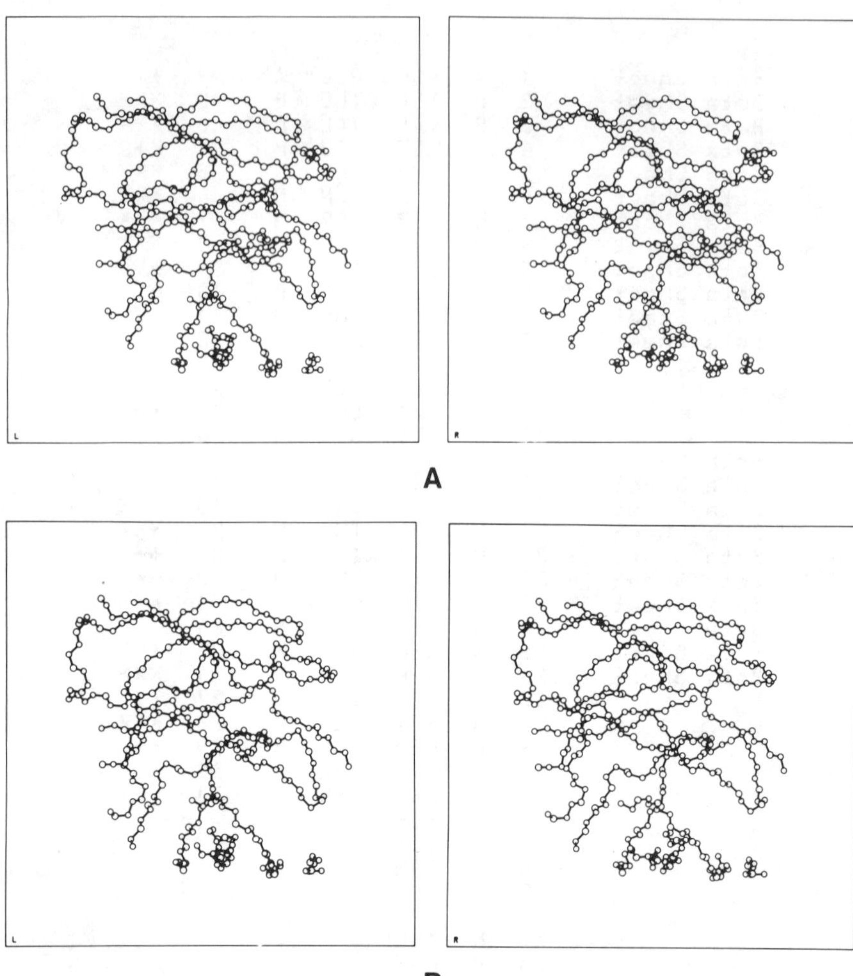

FIGURE 14. Ball and stick stereo of IgG J539 variable domain loop showing deletion. (**A**) Before modeling. (**B**) After modeling.

same sort of action that occurs when one is pulling on the faucet end of a hose which is laying out in one's garden.) This is a difficult task by most graphical procedures. It is much easier to leave the peptide backbone in the original beta strand conformation and modify the side chain assignments to close up the deletion. This has the effect of moving the deletions out onto the loop end. Such a sequence of modeling events is shown in FIGURE 12. During modeling the conformation of the peptide backbone at least four amino acids back from the deletion on either side is changed to bring the ends of the loop back into the proper distance and angles for the peptide bond.

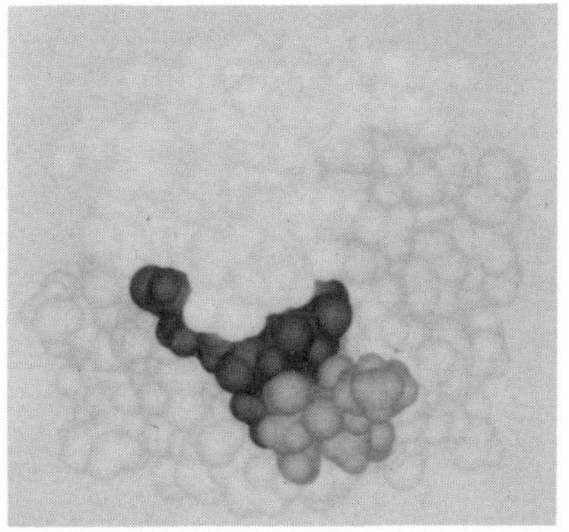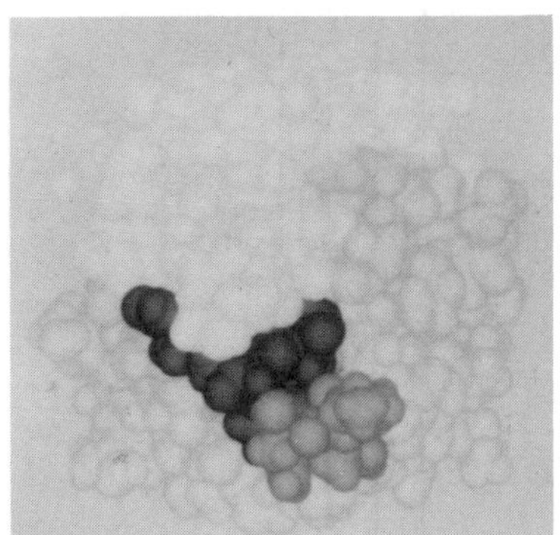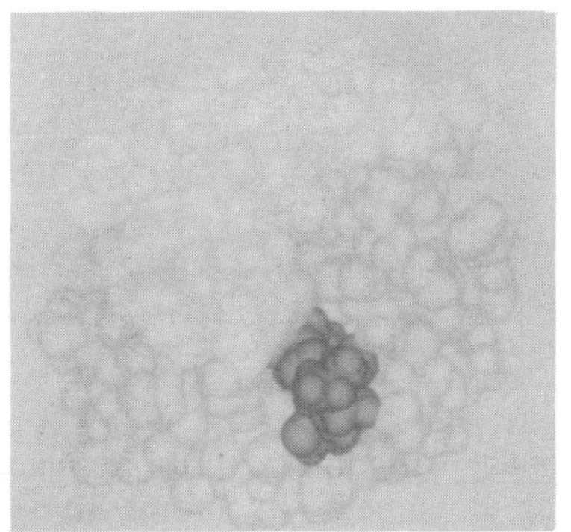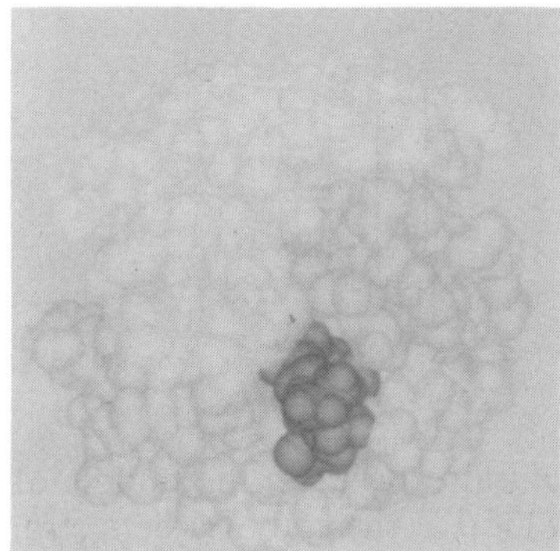

FIGURE 15. Space-filling stereo of IgG J539 variable domain loop showing deletion.

FIGURE 13 shows a portion of the alignment of the crystal structure immunoglobulin McPC–603 with the immunoglobulin J539. Notice that the homology is relatively strong and that the deletion occurs in a loop region between beta strands.

FIGURE 14 shows the ball and stick of the peptide backbone in the region of the deletion. Since the J539 specific side chains are not shown in this figure, the conformation of FIGURE 14A is effectively the conformation of McPC–603. FIGURE 14B is the conformation of the backbone with the deletion modeled out. FIGURE 15 shows the same structure states in space filling.

LOOP INSERTIONS

Insertions in the peptide chain during transformation of the crystal structure into the target structure are more difficult than deletions. Insertions which

```
A-B-C-D-E
| | | | \
| | | |  F              Before modeling
| | | |  |
| | | |  G
| | | | /
N-K-J-I-H


A-B-C-D-E
| | | | \
| | | |  F   H          During modeling
| | | |  \ /
| | | |  / \
| | | |  I   G
| | | | /
N-M-L-K-J


A-B-C-D-E-F
| | | | \
| | | |  G             After modeling
| | | |  |
| | | |  H
| | | | /
N-M-L-K-J-I
```

FIGURE 16. Insertion scheme, showing insertion of residues L and M.

would also occur in the middle of secondary structure features are placed in a loop at the end of a beta strand or an alpha helix. In contrast to deletions, an insertion in the loop region is characterized by having extra copies of amino acids added in the conformation of the loop. As shown in FIGURE 16, during modeling the amino acid H is a conformation copy of amino acid F but it is connected to amino acid I. Similarly amino acid G is a copy of I but connected to F. The modeling changes the conformation of at least four amino acids which are positioned back from the insertion junction. In the example case this will be between amino acids G and H.

FIGURE 17 is the match between factor B protein and chymotrypsin. Three amino acids will be inserted in the loop region between two beta strands. FIGURE 18 shows the ball and stick representation before and after modeling of this insertion. FIGURE 19 shows the same structure states in the space-filling mode.

Insertions which are longer than three or four amino acids are extremely difficult to model. FIGURE 20 shows a match in the central region of the chymotrypsin and factor B sequences; the homology is rather poor. The factor B protein has an insertion of nine amino acids. It is very difficult to determine how to position the conformation of these amino acids. FIGURE 21 shows another portion of the match where the homology is better, but there is an insertion of twelve amino acids. Continuing with the illustration in FIGURE

```
                      1   N   CYS                                              .
                      2   N   GLY                                              .
                      3   P   VAL                                              .
                      4   P   PRO                                              .
                      5   N   ALA                                              .
                      6   P   ILE                                              .
                      7   N   GLN                                              .
                      8   P   PRO                                              .
                     16   P   ILE   TRP   N   16    ++                         .
                     17   P   VAL   GLU   A   17    --                         .
                     18   N   ASN   HIS   N   18    +                          .
                     19   N   GLY   ARG   A   19    -                          .
                     20   A   GLU   LYS   A   20    +                          .
                     21   A   GLU   GLY   N   21    -                          .
                     22   N   ALA   THR   N   22    +                          .
                     23   P   VAL   ASP   A   23    --                         .
                     24   P   PRO   TYR   N   24    -                          .
                     25   N   GLY   HIS   N   25                               .
                     26   N   SER   LYS   A   26                               .
                     27   N   TRP   GLN   N   27    +                          .
Beta Sheet           28   P   PRO   PRO   P   28    ***                        .
Beta Sheet           29   N   TRP   TRP   N   29    ***                        .
Beta Sheet           30   N   GLN   GLN   N   30    ***                        .
Beta Sheet           31   P   VAL   ALA   N   31    ++                         .
Beta Sheet           32   N   SER   LYS   A   32                               .
Beta Sheet           33   P   LEU   ILE   P   33    ++                         .
Beta Sheet           34   N   GLN   SER   N   34    ++                         .
Beta Sheet           35   A   ASP   VAL   P   35    --                         .
                                    ILE   P  1036           1035   PRO         .
                                    ARG   A  1037           1036   ILE         .
                                    PRO   P  1038           1037   ARG         .
                     36   A   LYS   SER   N   36                               .
                     37   N   THR   LYS   A   37                               .
                     38   N   GLY   GLY   N   38    ***                        .
                     39   P   PHE   HIS   N   39    -                          .
                     40   N   HIS   GLU   A   40                               .
                     41   P   PHE   SER   N   41    -                          .
Beta Sheet           42   N   CYS   CYS   N   42    ***   **************.**
Beta Sheet           43   N   GLY   MET   P   43    ++                     . *
Beta Sheet           44   N   GLY   GLY   N   44    ***                    . *
Beta Sheet           45   N   SER   ALA   N   45    +                      . *
Beta Sheet           46   P   LEU   VAL   P   46    ++                     . *
Beta Sheet           47   P   ILE   VAL   P   47    ++                     . *
Beta Sheet           48   N   ASN   SER   N   48    ++                     . *
Beta Sheet           49   A   GLU   GLU   A   49    ***                    . *
Beta Sheet           50   N   ASN   TYR   N   50    +                      . *
Beta Sheet           51   N   TRP   PHE   P   51    ++                     . *
Beta Sheet           52   P   VAL   VAL   P   52    ***                    . *
```

FIGURE 17. The *N*-terminal portion of the alignment of chymotrypsin with factor B.

22 it can be seen that the factor B protein requires seven modeling actions to generate a model based on the chymotrypsin structure. When an insertion occurs the amino acid numbers are raised by 1000. Of the seven modeling actions six are insertions. The factor B protein is perhaps the most difficult protein we have pursued to date. When we constructed the models of the clotting factor proteins IXa, Xa and IIa, we thought that they were quite difficult because they had a number of insertions and deletions.[18,19] By later comparison with the factor B protein modeling experience, the clotting factor models were relatively easy.

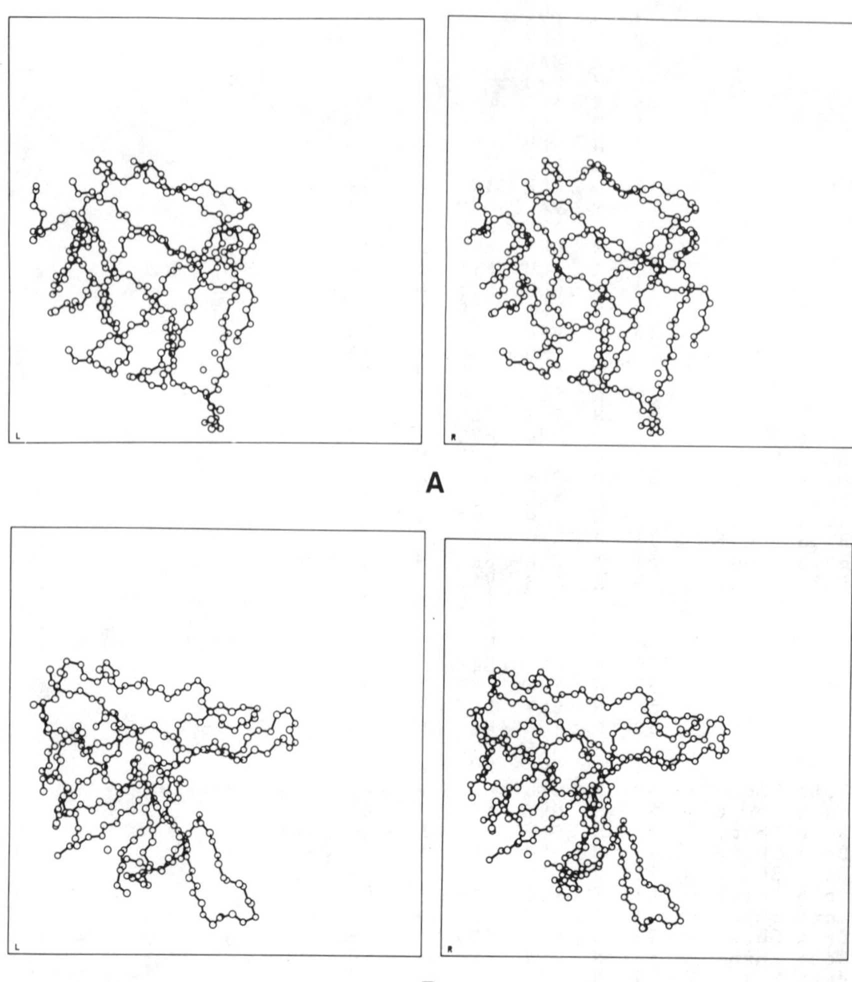

FIGURE 18. Ball and stick stereo of factor B protein loop showing insertion. (**A**) Before modeling. (**B**) After modeling.

GLOBAL MAPPING STABILITY

While developing the protein extension technique we tried to model interferon by matching the predicted alpha helical secondary structure of interferon to the helical structure of hemerythrin.[20] This was an interesting exercise but was totally unsuccessful as far as we were concerned. We felt that the sequence homologue of about fifteen percent was too low. Sternberg and Cohen thought otherwise.[21] When the homology between the crystal struc-

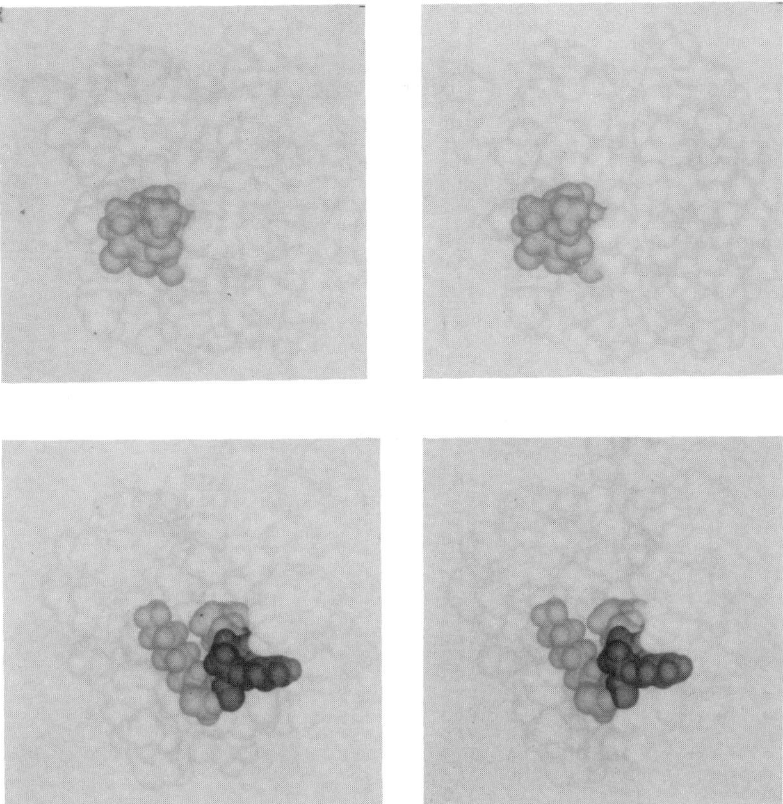

FIGURE 19. Space-filling stereo of factor B protein loop showing insertion.

ture protein and the target protein falls below about thirty percent it is diffi-cult to produce a model with the protein extension technique.

Only a portion of the factor B protein which has thirty percent homology with chymotrypsin was to be modeled. This was the *C*-terminal 274 amino which appeared to be a classic double beta barrel structure that was map-pable onto chymotrypsin. For the structure of the other domains of the factor B protein we can do very little, because at thirty percent homology with major insertions, the global stability of the mapping done is low. A stronger tree generation and tree evaluation program of the artificial intelligence type should be developed to explore this instability.

GRAPHICAL AND COMPUTATIONAL METHODS

The protein extension[22–24] technique as we have developed it differs from the attempts of others in that the flow of information from the mapping of the

```
                   129   A   ASP   ARG   A   129    +
                   130   P   PHE   LEU   P   130    ++
                   131   N   ALA   PRO   P   131    ++
                   132   N   ALA   PRO   P   132    ++
Beta Sheet         133   N   GLY   THR   N   133    +
Beta Sheet         134   N   THR   THR   N   134    ***
Beta Sheet         135   N   THR   THR   N   135    ***
Beta Sheet         136   N   CYS   CYS   N   136    ***       .....................
Beta Sheet         137   P   VAL   GLN   N   137    -                             .
Beta Sheet         138   N   THR   GLN   N   138    ++                            .
Beta Sheet         139   N   THR   GLN   N   139    ++                            .
Beta Sheet         140   N   GLY   LYS   A   140    -                             .
Beta Sheet         141   N   TRP   GLU   A   141    -                             .
                   142   N   GLY   GLU   A   142    -                             .
                   143   P   LEU   LEU   P   143    ***                           .
                   144   N   THR   LEU   P   144    -                             .
                   145   A   ARG   PRO   P   145    --                            .
                   146   N   TYR   ALA   N   146                                  .
                   147   N   ALA   GLN   N   147    +                             .
                   148   N   ASN   ASP   A   148                                  .
                   149   N   ALA   ILE   P   149    ++                            .
                   150   N   ASN   LYS   A   150                                  .
                   151   N   THR   ALA   N   151    +                             .
                   152   P   PRO   LEU   P   152    ++                            .
                               PHE   P  1153              1149   GLU             .
                               VAL   P  1154              1150   LYS             .
                               SER   N  1155              1151   LYS             .
                               GLU   A  1156              1152   LEU             .
                               GLU   A  1157              1153   PHE             .
                               GLU   A  1158              1154   VAL             .
                               LYS   A  1159              1155   SER             .
                               LYS   A  1160              1156   GLU             .
                               LEU   P  1161              1157   GLU             .
                   153   A   ASP   THR   N   153                                  .
                   154   A   ARG   ARG   A   154    ***                           .
Beta Sheet         155   P   LEU   LYS   A   155    --                            .
Beta Sheet         156   N   GLN   GLU   A   156                                  .
Beta Sheet         157   N   GLN   VAL   P   157    -                             .
Beta Sheet         158   N   ALA   TYR   N   158                                  .
Beta Sheet         159   N   SER   ILE   P   159    -                             .
Beta Sheet         160   P   LEU   LYS   A   160    --                            .
Beta Sheet         161   P   PRO   ASN   N   161    -                             .
Beta Sheet         162   P   LEU   GLY   N   162    ++                            .
                   163   P   LEU   ASP   A   163    --                            .
                   164   N   SER   LYS   A   164                                  .
                   165   N   ASN   LYS   A   165                                  .
                   166   N   THR   GLY   N   166    +                             .
                   167   N   ASN   SER   N   167    ++                            .
                   168   N   CYS   CYS   N   168    ***   ******************.**
                   169   A   LYS   GLU   A   169    +                        . *
```

FIGURE 20. One of the central portions of the alignment of chymotrypsin with factor B.

sequences to the photography of the color stereo space filling images[25-28] is very smooth. In cases where the length of the protein family happens to be conserved, the whole family can be generated automatically after the multiple sequence alignment is made.

The EDIT-SUBSTITUTION-SEQUENCE function in the program EDISEQ generates a command file that will transform the crystal structure into the target sequence. The two columns of FIGURE 23 show a portion of a typical command file. NIH225 is the bovine phospholipase a2 and BB2 is one of the snake neurotoxins. This command file can be automatically generated because the sequence editor has all the information about the trans-

Beta Sheet										
	164	N	SER	LYS	A	164				.
	165	N	ASN	LYS	A	165				.
	166	N	THR	GLY	N	166	+			.
	167	N	ASN	SER	N	167	++			.
	168	N	CYS	CYS	N	168	***			***************** .**
	169	A	LYS	GLU	A	169	+			. *
	170	A	LYS	ARG	A	170	++			. *
	171	N	TYR	ASP	A	171	+			. *
	172	N	TRP	ALA	N	172	++			. *
	173	N	GLY	GLN	N	173	+			. *
	174	N	THR	TYR	N	174	+			. *
	175	A	LYS	ALA	N	175	-			. *
				PRO	P	1176		1170	LYS	. *
				GLY	N	1177		1171	ASP	. *
				TYR	N	1178		1172	ILE	. *
				ASP	A	1179		1173	SER	. *
				LYS	A	1180		1174	GLU	. *
				VAL	P	1181		1175	VAL	. *
				LYS	A	1182		1176	PRO	. *
				ASP	A	1183		1177	GLY	. *
				ILE	P	1184		1178	TYR	. *
				SER	N	1185		1179	ASP	. *
				GLU	A	1186		1180	LYS	. *
				VAL	P	1187		1181	VAL	. *
	176	P	ILE	VAL	P	176	++			. *
	177	A	LYS	THR	N	177				. *
	178	A	ASP	PRO	P	178	--			. *
Beta Sheet	179	N	ALA	ARG	A	179	-			. *
Beta Sheet	180	P	MET	PHE	P	180	++			. *
Beta Sheet	181	P	ILE	LEU	P	181	++			. *
Beta Sheet	182	N	CYS	CYS	N	182	***			***************** .**
Beta Sheet	183	N	ALA	THR	N	183	+			.
Beta Sheet	184	N	GLY	GLY	N	184	***			.
Beta Sheet	185	N	ALA	GLY	N	185	++			.
Beta Sheet	186	N	SER	VAL	P	186	-			.
	187	N	GLY	SER	N	187	+			.
				PRO	P	1188		1186	ALA	.
				TYR	N	1189		1187	ASP	.
				ALA	N	1190		1188	PRO	.
				ASP	A	1191		1189	TYR	.
	188	P	VAL	PRO	P	188	++			.
	189	N	SER	ASN	N	189	++			.
	190	N	SER	THR	N	190	++			.
	191	N	CYS	CYS	N	191	***			....
	192	P	MET	ARG	A	192	--			.
Beta Sheet	193	N	GLY	GLY	N	193	***			.
Beta Sheet	194	A	ASP	ASP	A	194	***			.
Beta Sheet	195	N	SER	SER	N	195	***			.
Beta Sheet	196	N	GLY	GLY	N	196	***			.
Beta Sheet	197	N	GLY	GLY	N	197	***			.
Beta Sheet	198	P	PRO	PRO	P	198	***			.
Beta Sheet	199	P	LEU	LEU	P	199	***			.
Beta Sheet	200	P	VAL	ILE	P	200	++			.
Beta Sheet	201	N	CYS	VAL	P	201	-			....
	202	A	LYS	HIS	N	202	+			.
	203	A	LYS	LYS	A	203	***			.
	204	N	ASN	ARG	A	204				.
	205	N	GLY	SER	N	205	+			.
	206	N	ALA	ARG	A	206	-			.
	207	N	TRP	PHE	P	207	++			.
	208	N	THR	ILE	P	208	-			.
	209	P	LEU	GLN	N	209	-			.
	210	P	VAL	VAL	P	210	***			.
	211	N	GLY	GLY	N	211	***			.
Beta Sheet	212	P	ILE	VAL	P	212	++			.
Beta Sheet	213	P	VAL	ILE	P	213	++			.
Beta Sheet	214	N	SER	SER	N	214	***			.
Beta Sheet	215	N	TRP	TRP	N	215	***			.
Beta Sheet	216	N	GLY	GLY	N	216	***			.
Beta Sheet	217	N	SER	VAL	P	217	-			.
Beta Sheet	218	N	SER	VAL	P	218	-			.
	219	N	THR	ASP	A	219				.
	220	N	CYS	VAL	P	220	-			....
	221	N	SER	CYS	N	221	++			.
	222	N	THR	LYS	A	222				.
Beta Sheet	223	N	SER	ASN	N	223	++			.
Beta Sheet	224	N	THR	GLN	N	224	++			.
Beta Sheet	225	P	PRO	LYS	A	225	--			.
Beta Sheet	226	N	GLY	ARG	A	226	-			.

FIGURE 21. Another of the central portions of the alignment of chymotrypsin with factor B protein.

```
 File Name  |  Lower,Upper  |  Done  |  Ease  |
 ---------- | ------------- | ------ | ------ |
    1PB     |  1037,1035    |        |        |
 ---------- | ------------- | ------ | ------ |
    2PB     |  1065,1061    |        |        |
 ---------- | ------------- | ------ | ------ |
    3PB     |  1129,1125    |        |        |
 ---------- | ------------- | ------ | ------ |
    4PB     |  1157,1149    |        |        |
 ---------- | ------------- | ------ | ------ |
    5PB     |  1181,1170    |        |        |
 ---------- | ------------- | ------ | ------ |
    6PB     |  1189,1186    |        |        |
 ---------- | ------------- | ------ | ------ |
    7PB     |   245,1231    |        |        |
 ---------- | ------------- | ------ | ------ |
```

FIGURE 22. Table of file actions required to model factor B protein. Match value = 208. Match percent = 30.14.

formation. When executed the command file calls into operation various programs and functions in the molecular modeling system and supplies these functions with the appropriate file names and parameters. From a programming viewpoint, generating command files which call various programs is much easier than including all the functions in one giant program. Much of the modeling of macromolecular assemblies is done by hierarchies of command files.[29] On the DECsystem-10, we can have values substituted into symbolic variables as they are executed.

ANALYSIS OF EXTENDED MODELS

The amino acids on the surface of a protein are generally less constrained than those which are inside the protein or which have side chains that point into the center of the protein. Constraints on surface side chains exist when these amino acids play a role producing the complimentary surface between dimers or an interface between the protein and a membrane. In this regard one of the types of images which has proven to be useful and thus popular is myoglobin. In FIGURE 24 the differences between the crystal structure and the target structure are illustrated. It can be seen that the myoglobins are conservative and the differences on the face of the protein where the heme is exposed are very few. The immunoglobulins are less conserved and typically two random immunoglobulins will have only fifty percent homology.

In FIGURE 25A the same difference coloring is shown for the variable domains of the immunoglobulin J539 modeled from McPC–603. FIGURE 25B shows the portions of McPC–603 variable domain which will be deleted in making the model of J539. Again, the difference coloring for the factor B protein with respect to chymotrypsin is shown in FIGURE 26.

```
; INPUT FILE =NIH225          GLU
; Output file =BB2            40,46,71
.GOTO 'A                      GLY
STEP1::                       15
.PRO NIH225.X?<055>           HIS
.SOS NIH225.XR                17,114
SNT   N   EF                  LEU
ET                            117
.PRO NIH225.X?<555>           LYS
.DEL ?BB2.X?                  87,115
.RUN XRAY[14,14]              MET
NEW                           20
NIH225                        PHE
ACIDR                         63
I                             SER
1,10000                       7,72
  1000                        THR
PICK                          80,119
                              TYR
·1                            122
;Insertions
1114                          SBB2
                              SAVE
                              SBB2
CSAVE                         Substituted structure for BB2
SAVE
UBB2                          END
Upper residues of BB2         STEP3::
                              .DEL ABB2.XR
NEW                           .RUN RESSUB[14,14]
NIH225                        BLOCK-UPPER
.RUN XRAY[14,14]              NEW
BLOCK                         SBB2
DNEW                          ANNEAL
NIH225                        N
UBB2                          SBB2
SAVE                          ABB2
CBB2                          FILEMODE
; Deletions and insertions
END                           SAVE
STEP2::                       ABB2
.DEL SBB2.XE                  After annealing of SBB2
.RUN RESSUB[14,14]
NEW                           END
CBB2                          .RUN RESSUB[14,14]
PEP-S                         NEW
;Normal substitutions         ABB2
ALA                           BREAK-SULFUR-BONDS
12,109                        INSERTION
ARG                           ;Put insertion reconnections h
6,53                          R, 113,1114,1114
[the figure continues in the
next column]
```

FIGURE 23. A section of the code generated by EDISEQ to transform bovine phospholipase a2 into XXX neurotoxin.

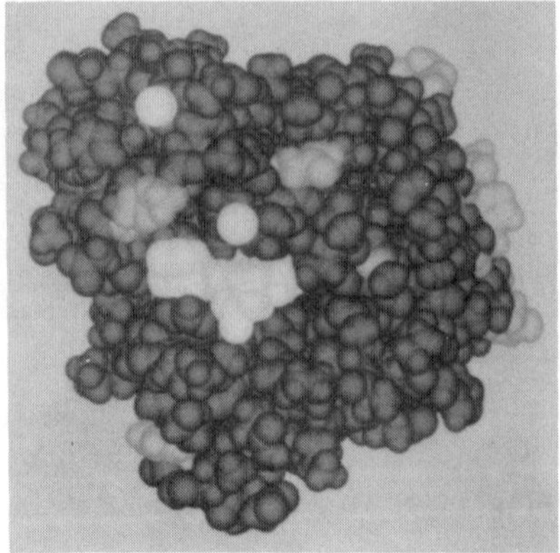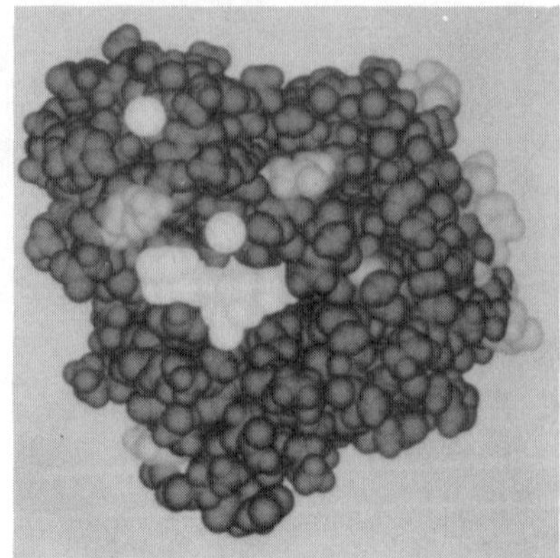

FIGURE 24. Space-filling stereo of minke whale myoglobin showing differences from sperm whale myoglobin.

FAMILIES THAT HAVE BEEN EXTENDED

We have developed the protein extension procedures as a method for building atomic level models of protein families. Over the three years of its development we have constructed models for approximately 300 proteins. The flexible and in some cases the automatic construction of models is the key to this large number. We have built models for the following families: lysozymes, myoglobins, cytochromes, ovomucoids, renin, immunoglobulins, snake toxins, insulins, serine proteases, dehydrogenases, and dihydrofolate reductases.

The families of models were constructed in response to the needs of scientists at the NIH and at various institutions in America and abroad. The myoglobin family was constructed with Jay Berzofsky in the National Cancer

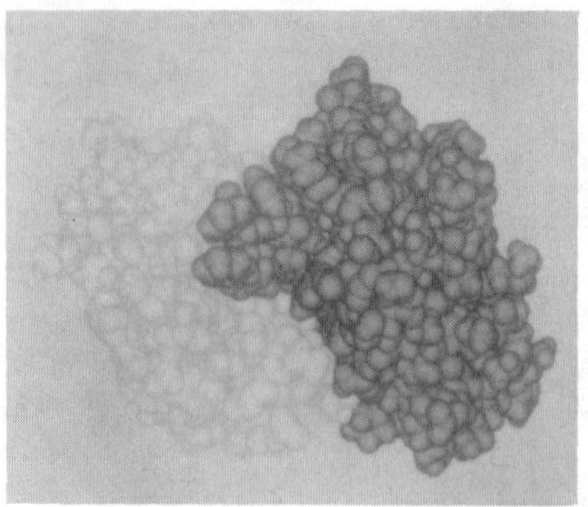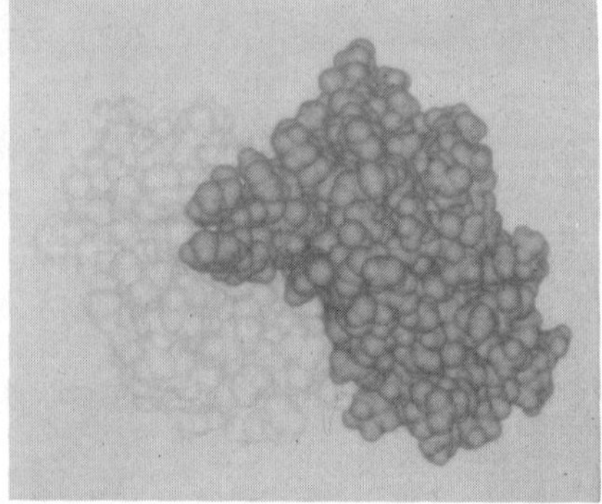

FIGURE 25A. Space-filling stereo of IgG J539 variable domain showing amino acids to be deleted to form IgG McPC–603 variable domain.

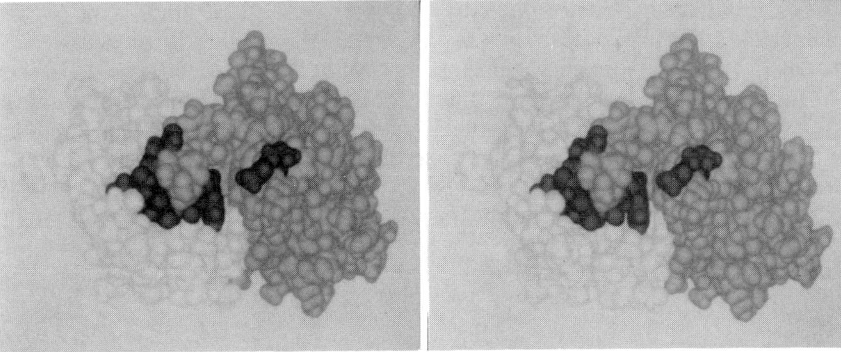

FIGURE 25B. Space-filling stereo of IgG McPC–603 variable domain showing amino acids to be deleted to form IgG J539 variable domain.

Institute at the NIH[30–32] in order to identify how monoclonal antibodies were specifically binding to the myoglobins. The ovomucoids were constructed for John Markley who does NMR experiments at Purdue University. Markley used the single amino acid differences between members of the family to produce discretely different NMR signals. The snake toxins were developed for Trevor Payne at Sandoz in Basel. The protein extension method was used to correct one sequence version of mouse lactate dehydrogenase.[33–35] Several of us (MP, CM and RJF) have produced models of a variety of immunoglobulins.[37–39] Another grouping (DHB, BF, BCF and RJF) has produced the serine protease models.[19,20] Preliminary models of bovine rhodopsin which include only the helical segments have been produced.[40] The modeling system can also be used to build models of DNA.[41]

The reasons for which individual scientists build protein models are as varied as the protein families themselves are random. This makes generalizing

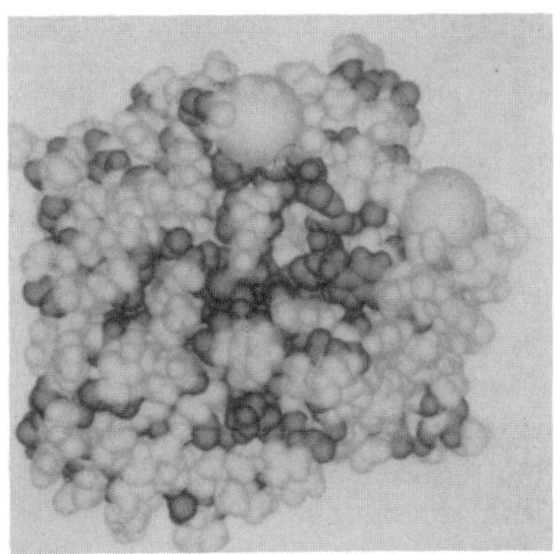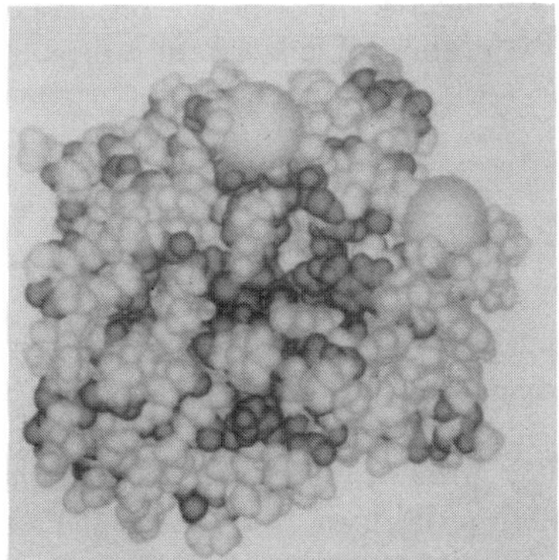

FIGURE 26. Space-filling stereo of factor B protein showing differences from chymotrypsin.

to such a diverse audience difficult. However, it is rather important at this particular time in the development of our understanding of protein structure to get a three-dimensional sense of amino acid sequence conservation within and across families of proteins. There is still no clear understanding of which particular aspects of protein sequence contribute to protein folding. The exterior surfaces of globular proteins for which a crystal structure has been found appear to be quite variable with respect to amino acid sequence. Since there are twenty different amino acids the combinatorics of protein mutation for any large protein are vast. One hardly knows how to go about rationally mutating these large proteins.[41] This remains a major problem that will affect all protein modeling experiments.

ENERGY MINIMIZATION OF EXTENDED MODELS

The models created by the protein extension technique are syntactically correct. The removal of van der Waals contacts is a crude form of energy minimization. Preliminary experiments with various energy minimization programs are in progress at the writing of this paper. One of the main drawbacks to the use of these energy minimization programs is that they require very long times for execution. Williams Carlson, working at the Massachusetts General Hospital, has energy minimized the model of renin constructed by our technique using the program CHARM from Martin Karplus and his co-workers at Harvard University.

Carlson used about 1000 hours of DEC Vax-780 time in doing this minimization. The starting crystallographic structure and the renin were minimized together. During the first phases of the minimization it became clear that while the energy of the acid protease was continuing to fall from very high positive energy levels to negative energy levels, the energy level of the renin model stopped falling once the main surface van der Waals contacts had been eliminated. We looked at the structure in detail and found that we had made a mistake in making the mapping between the three-dimensional structure in the active site and the sequence of the renin. We rebuilt the renin with a different active site mapping and then reminimized. The energy levels of the renin structure fell to the same level.

CONCLUSION

X-ray crystallography is currently the only way of obtaining very accurate structures of proteins. Crystallography as a process is rate limited at several stages (*i.e.*, obtaining materials, purification, crystallization, derivative production, map solving and structure refinement). Where possible, protein extension applies the results of known crystallographic information to build other protein models. We have shown that some models are easy to construct and some are difficult. Probably some models are accurate representations of the protein, but most are at best first approximations of the three-dimensional structure of the protein. While poor models can of course be misleading, even a poor model will hopefully cause the user/biochemist/geneticist to think about the problem. Thus the use of protein extended models is perfectly valid.

REFERENCES

1. THOMAS, K. A. & A. N. SCHECHTER. 1980. Protein folding. Biol. Reg. Dev. **2**: 43–100.
2. KABSCH, W & C. SANDER. 1983. How good are predictions of protein secondary structure? F.E.B.S. Lett. **155**: 179–182.
3. NEMETHY, G & H. A. SCHERAGA. 1977. Protein folding. Quart. Rev. Biophys. **10**: 239–352.
4. RICHARDSON, J. 1981. The anatomy and taxonomy of protein structure. Adv. Protein Chem. **34**: 167–339.
5. DAYHOFF, M. O. & W. C. BARKER. 1972. Atlas of Protein Sequence and Structure, vol. 5. National Biomedical Research Foundation. Washington, D. C.
6. BARKER, W. C. Private communication.
7. CHOU, P. Y. & G. D. FASMAN. 1974. Biochemistry **13**: 221–245.
8. SCHULZ, G. E., C. D. BARRY, J. FRIEDMAN, P. Y. CHOU, G. D. FASMAN, A. V. FINKELSTEIN, V. I. LIM, O. B. PTITSYN, E. A. KABAT, T. T. WU, M. LEVITT, B. ROBSON & K. NAGANO. 1974. Nature **250**: 140–142.
9. EISENBERG, D., R. M. WEISS & T. C. TERWILLIGER. 1982. The helical hydrophobic moment: A measure of the amphiphilicity of a helix. Nature **299**: 371–374.
10. KYTE, J. & R. DOOLITTLE. 1982. A simple method for displaying the hydrophobic character of a protein. J. Mol. Biol. **157**: 105–132.
11. ROBSON, B. & D. J. OSGUTHORPE. 1979. Refined models for computer simulation of protein folding. J. Mol. Biol. **132**: 19–51.
12. HOPP, T. P. & K. R. WOODS. 1981. Prediction of protein antigenic determinants from amino acid sequences. Proc. Natl. Acad. Sci. U.S.A. **78**: 3824–3828.
13. CRAIK, C. S., S. SPRANG, R. FLETTERICK & W. J. RUTTER. 1982. Intron-exon splice junctions map at protein surfaces. Nature **299**: 180–182.
14. CRAIK, C. S., W. J. RUTTER & R. FLETTERICK. 1983. Splice junctions: Association with variation in protein structure. Science **220**: 1125–1129.
15. FELDMANN, R. J. 1983. Directions in macromolecular structure representation and display. *In* Computer Applications in Chemistry. S. R. Heller & R. Potenzone, Jr., Eds.: 9–18. Elsevier. Amsterdam.
16. WILBUR, W. J. & D. J. LIPMAN. 1983. Rapid similarity searches of nucleic acid and protein data banks. Proc. Natl. Acad. Sci. U.S.A. **80**: 726–730.
17. DICKERSON, R. E. & I. GEIS. 1969. The Structure and Action of Proteins. Harper and Row. New York. pp. 16–17.
18. FURIE, B., D. H. BING, R. J. FELDMANN, D. J. ROBINSON, J. P. BURNIER & B. C. FURIE. 1982. Computer-generated models of blood coagulation factor Xa, factor IXa, and thrombin based upon structural homology with other serine proteases. J. Biol. Chem. **257**: 3875–3883.
19. BING, D. H., R. LAURA, D. J. ROBISON, B. FURIE, B. C. FURIE & R. J. FELDMANN. A computer-generated three-dimensional model of the B chain of bovine alpha-thrombin. Ann. N.Y. Acad. Sci. **370**: 496–510.
20. WEBER, P. C. & F. R. SALEMME. 1980. Structural and functional diversity in 4-alpha-helical proteins. Nature **287**: 82–84.

21. STERNBERG, M. J. E. & F. E. COHEN. 1982. Prediction of the secondary and tertiary structures of interferon from four homologous amino acid sequences. Int. J. Biol. Macromol. **4**: 137–144.

22. MOEWS, P. C. & J. R. KNOX. 1979. Predicted secondary structures of four penicillin beta-lactamases and a comparison with two lysozymes. Int. J. Petpide Protein Res. **13**: 385–393.

23. BEDARKAR, S., W. G. TURNELL & T. L. BLUNDELL. 1977. Relaxin has conformational homology with insulin. Nature **270**: 449–451.

24. JORNVALL, H., A. CARLSTROM, T. PETTERSSON, B. JACOBSSON, M. PERSSON & V. MUTT. 1981. Structural homologies between prealbumin, gastrointestinal prohormones and other proteins. Nature **291**: 261–263.

25. FELDMANN, R. J. & D. H. BING. 1980. Teaching Aids for Macromolecular Structure (TAMS). Taylor Merchant, Inc. New York. 140 pp. and 116 stereo slides and student unit of 49 stereo slides.

26. FELDMANN, R. J., D. H. BING, B. C. FURIE & B. FURIE. 1978. Interactive computer surface graphic approach to the study of the active site of bovine trypsin. Proc. Natl. Acad. Sci. U.S.A. **75**: 5409–5412.

27. PORTER, T. K. 1978. Spherical shading. Computer Graphics **12**: 282–285.

28. PORTER, T. K. 1979. The shaded surface display of large molecules. Computer Graphics **13**: 234–236.

29. FELDMANN, R. J. 1976. The design of computing systems for molecular modeling. Ann. Rev. Biophys. Bioeng. **5**: 477–510.

30. BERZOFSKY, J. A., G. K. BUCKENMEYER, G. HICKS, F. R. N. GURD, R. J. FELDMANN & J. MINNA. 1982. Topographic antigenic determinants, recognized by monoclonal antibodies to sperm whale myoglobin. J. Biol. Chem. **257**: 3189–3198.

31. BERZOFSKY, J. A., G. K. BUCKENMEYER, G. HICKS, D. J. KILLION, I. BERKOWER, Y. KOHNO, M. A. FLANAGAN, R. J. FELDMANN, J. MINNA & F. R. N. GURD. 1983. Topographic antigenic determinants recognized by monoclonal antibodies to myoglobin. *In* Protein Conformation as an Immunological Signal. F. Celafa, V. N. Schumaker & E. E. Sercarz, Eds. Plenum. New York.

32. BERKOWER, I., G. K. BUCKENMEYER, F. R. N. GURD & J. A. BERZOFSKY. 1982. A possible immunodominant epitope recognized by murine T lymphocytes immune to different myoglobins. Proc. Natl. Acad. Sci. U.S.A. **79**: 4723–4727.

33. LI, S. I. & R. J. FELDMANN. 1983. Molecular features and immunological properties of lactate. M. Okabe & Y. C. Pan. J. Biol. Chem. **258**: 7017–7028.

34. LI, S. I., W. M. FITCH, Y. -C. E. PAN & R. J. FELDMANN. 1983. Evolutionary relationships of vertebrate lactate dehydrogenase isozymes A4 (muscle), B4 (heart), and C4 (testis). J. Biol. Chem. **258**: 7029–7032.

35. OKABE, M. & Y. C. PAN. 1983. J. Biol. Chem. **258**: 7017–7028.

36. FELDMANN, R. J., M. POTTER & C. P. J. GLAUDEMANS. 1981. A hypothetical space-filling model of the V-regions of the galactan-binding myeloma immunoglobulin J539. Mol. Immun. **18**: 683–698.

37. PAWLITA, N., E. MUSHINSKI, R. J. FELDMANN & M. POTTER. 1981. A monoclonal antibody that defines an idiotope with two subsites in galactan-binding myeloma proteins. J. Exp. Med. **154**: 1946–1956.

38. GUTMAN, G. A. 1981. Genetic and structural studies on rat kappa chain allotypes. Transplant. Proc. **13**: 1483–1488.

39. HARGRAVE, P. A., J. H. MCDOWELL, E. C. SIEMIATKOWSKI-JUSZCZAK, S. -L. FONG, H. KUHN, J. K. WANG, D. R. CURTIS, J. K. MOHANA RAO, P. ARGOS & R. J. FELDMANN. 1982. The carboxyl-terminal one-third of bovine rhodopsin: Its structure and function. Vision **22**: 1429–1438.

40. O'Neill, M. C., K. Amass & B. DeCrombrugghe. 1981. Moledular model of the DNA intercalation site for cyclic AMP receptor protein. Proc. Natl. Acad. Sci. U.S.A. **78**: 2213–2217.
41. Ulmer, K. M. 1983. Protein engineering. Science **219**: 666–671.

Protein Structure and Function by Comparative Model Building

JONATHAN GREER

Physical Biochemistry Laboratory
Computer-Assisted Molecular Design
Abbott Laboratories
Abbott Park, North Chicago, Illinois 60064

INTRODUCTION

Comparative modeling methods have been used to extend the experimentally determined three-dimensional structures of proteins to new molecules whose structure is closely related. Such techniques have been applied to derive model structures for α-lactalbumin,[1] α-lytic protease,[2] *Streptomyces* trypsin-like protein,[3] Ca^{++} binding proteins,[4] haptoglobin,[5] serine proteases,[6] including blood clotting factor Xa[7] and very recently renin,[8] a member of the acid protease family. Thus, comparative model building has been widely used for producing tentative structures of biologically interesting and important molecules.

In this study, we employ comparative modeling methods to begin exploring the nature of specificity between enzymes and their particular substrates. Initial structures for several serine proteases are derived.[5-7] Suitable peptides from the known macromolecular substrates of these enzymes are also modeled onto the enzyme active site[7] and their properties in relation to their respective enzymes are analyzed and compared. The ultimate goal is to gain a deeper and more detailed understanding of the molecular basis of enzyme-substrate and protein-ligand recognition and specificity.

METHODS

Comparative Modeling Method

A detailed description of the comparative modeling methods for the serine proteases used in this work has been published.[6] Briefly, by comparing the experimentally known structures of chymotrypsin,[9,10] trypsin,[11,12] and elastase,[13,14] the three-dimensional structures of the serine proteases can be parsed into structurally conserved regions (SCRs) and variable regions (VRs) (see FIG. 1). Sequence homology among these proteins lies almost exclusively in the SCRs (see Figure 2 of ref. 6).

To model a "new" serine protease, the sequence is aligned using the strong

FIGURE 1. An α-carbon plot of the structurally conserved regions (*dotted lines*) of chymotrypsin together with the different variable regions for each of the three serine protease structures: chymotrypsin (*dotted lines*), trypsin (*dashed lines*), and elastase (*solid lines*). The three conformations at a variable region are considered when modeling the structure of a "new" protein at that loop.

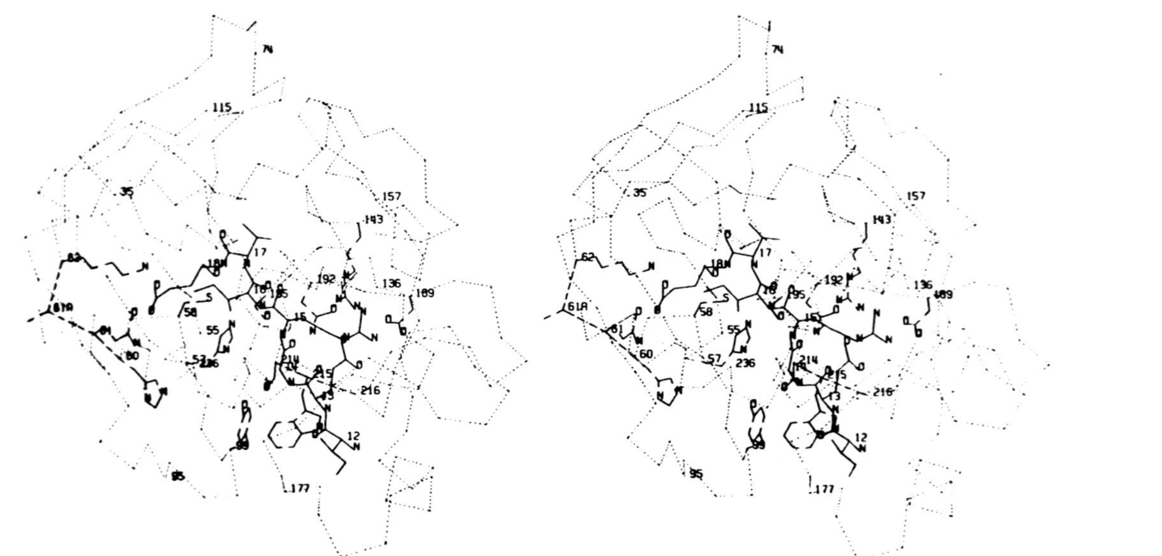

FIGURE 2. Model structure for the factor Xa-prothrombin complex in stereo. The α-carbon plot of factor Xa is shown (*dotted lines*) with the side chains (*dashed lines*) that are interacting with the prothrombin substrate (*solid lines*). Residue Arg P_1 (labeled 15) forms a salt bridge with Asp 189 of factor Xa. Glu P_3 (labeled 13) forms a bridge with Arg 143 and Glu P_3' (labeled 18) forms a bridge with Lys 62. Any larger side chain than Gly at site P_2 (labeled 14) would collide with Tyr 99 of factor Xa.

homology in the SCRs. The coordinates for the main chain in the SCRs are taken from any one of the known structures (since they are almost identical, see Figure 1 of ref. 6). The side chains are "mutated" to fit the sequence of the new protein.

Modeling the VRs is much more challenging. For each VR, the various conformations found amongst the known structures (FIG. 1) are examined as to length (see Table 3 of ref. 6) and residue character. If one of the known conformations fits the new sequence, its main chain coordinates are used directly with suitable replacement of side chains. Otherwise, modeling using energetics is necessary to achieve a reasonable tentative conformation for the respective VR. Modeling studies on eight serine protease sequences[6,7] show that more than 50% of the VRs can be modeled directly from the known structures whereas less than 5% fall in a class of large additions where modeling is not practical in the near future.

Modeling the Serine Proteases and Their Substrates

TABLE 1 lists the substrates and enzymes used in this study. The two serine proteases, blood clotting factors Xa and IXa, were sequenced by Titani *et al.*[15] and Katayama *et al.*,[16] respectively. The sequence alignments used for these two proteins are given in Figure 3 of ref. 6. The SCRs were modeled from elastase and the VRs as shown in TABLE 2. In the case of factor Xa (see ref. 7) three loops cannot be built without further energy analysis; however, these three VRs are distant from the active site region and thus do not concern us in this study. Factor IXa is closely related but not identical to factor Xa in the size of its VRs as can be seen from Figure 3 and Table 3 of ref. 6 and from TABLE 2. In factor IXa, only two loops have no direct known structure as a model. One of them, the VR at 36–38, lies just on the border of the substrate binding pocket. The other is distant from the active site region, as it was in factor Xa.

The substrate peptides, consisting of residues $P_4,..., P_1, P_1',..., P_3'$ (using the standard substrate nomenclature[17]), were modeled after the crystallographically determined conformations of bovine pancreatic trypsin inhibitor,[11] of soybean trypsin inhibitor,[18,19] and of di- and tripeptide chloromethyl ketones bound to γ-chymotrypsin[20,21] and to *Streptomyces griseus* protease B.[22] All these inhibitors have a common main chain conformation between residues P_3 and P_3' (see ref. 7 for a more complete discussion of this). The side chain positions for residues P_3 and P_3' as well as the position of residue

TABLE 1. Substrates and Enzymes Used

Substrate	Enzyme
prothrombin	factor Xa
factor X	factor IXa
trypsinogen	trypsin

TABLE 2. Model Structures for the VRs in Factors Xa and IXa

VR	Factor Xa		Factor IXa	
	Model	Residues built	Model	Residues built
23–25	elastase	—	elastase	—
36–38	chymotrypsin	33–40	deletion	(at 37)
59–62	elastase	—	elastase	—
72–80	deletion	(at 76)	elastase	—
97–101	trypsin	95–102	elastase	—
116	deletion	(at 116)	elastase	—
124–133	addition	(at 131)	addition	(at 131)
146–151	elastase	—	elastase	—
166–179	trypsin	164–180	trypsin	164–180
185–187	trypsin	184–188	trypsin	184–188
203–206	elastase	—	elastase	—
217–224	chymotrypsin	216–226	chymotrypsin	216–226

P_4 were determined by modeling the substrate on the respective enzyme as previously described.[7]

Surface Representations and Electrostatic Potential Maps

The solvent exclusion surfaces of the substrates and enzymes were calculated by the method of Connolly[23,24] using a program obtained from him.

The electrostatic potential maps were calculated as follows: a point unit positive charge was placed on the solvent-accessible surface and the electrostatic potential calculated from Coulomb's Law as

$$E_{el} = \sum_i \frac{q_i}{r_i}$$

where q_i is the charge on the ith atom and r_i is the distance from the point positive charge to this atom. The charges on the atoms were taken from Kollman and his co-workers.[25] A unit dielectric constant is used here because we are interested in the interactions between enzyme and substrate at the complex interface from which water should be excluded. The major conclusions of this work are not changed if a dielectric constant $= r_i$ is used as reported by many workers.[26-28]

RESULTS AND DISCUSSION

The Structures of the Enzyme-Substrate Complexes

The details of the proposed enzyme-substrate complex for factor Xa and prothrombin have previously been described[7] (FIG. 2). The cleaved peptide bond lies between residues P_1 and P_1' in prothrombin (see TABLE 3). Residue P_1 is an Arg and forms a salt bridge with Asp 189 of factor Xa. This interac-

TABLE 3. Sequence of Cleaved Peptide in Substrate

	P_4[a]	P_3	P_2	P_1	P_1'	P_2'	P_3'
Substrate	12[b]	13	14	15	16	17	18
Prothrombin	Ile	Glu	Gly	Arg	Ile	Val	Glu
Factor X	Gln	Val	Val	Arg	Ile	Val	Gly
Trypsinogen	Asp	Asp	Asp	Lys	Ile	Val	Gly

[a] Nomenclature as in ref. 17.

[b] This residue numbering corresponds to that for chymotrypsinogen which is often used for comparing serine protease sequences. These numbers are used to label the residues of the substrate peptides in Figs. 2–9.

tion is typical of many serine proteases and is the molecular basis of the primary specificity shown by these enzymes, including trypsin, to cleave only after Lys or Arg residues.[29] The main chain of residues P_2 and P_3 forms an antiparallel β-sheet with residues 216–218 of the factor Xa enzyme. This feature appears to be common to all the inhibitor structures examined experimentally and is assumed in all the complexes modeled in this study. Residue P_2 is a Gly and thus has no side chain; any larger side chain at this point would collide with the phenolic hydroxyl of Tyr 99 on factor Xa. The glutamate side chains at P_3 and P_3' form specific salt bridges with Arg 143 and Lys 62 on factor Xa, respectively. Finally, Ile P_4, Ile P_1', and Val P_2' all lie in hydrophobic regions of the molecule. Thus, to sum up, in addition to the Arg-Asp salt bridge conferring the primary specificity, there are two other salt bridges, a stereogeometric requirement for a Gly, and numerous hydrophobic interactions.

The complex between blood clotting factor X as a substrate and its activating enzyme factor IXa was also examined in detail (FIG. 3). The primary specificity interaction is the same as above: Arg P_1 of factor X with Asp 189 of factor IXa. The side chains of Val P_2 and Val P_3 appear to make few close interactions, nor do they seem to lie in a particularly hydrophobic environment. On the other hand, Gln P_4 has the possibility of making several hydrogen bonds with the hydroxyl side chains of Thr 172 and Ser 175. On the other side of the cleaved bond, the relatively invariant Ile P_1' and Val P_2' lie in conserved hydrophobic pockets on factor IXa. Residue P_3' lies immediately adjacent to the VR at positions 36–38, which cannot be modeled in factor IXa without detailed energetic analysis since no experimentally known conformation exists for this sequence (see METHODS and TABLE 2). However, the occurrence of Gly P_3' in factor X avoids the need to model this VR accurately because it has no side chain to fit to the enzyme as there was in the prothrombin-factor Xa complex (see above and ref. 7 for a detailed discussion). Therefore, the major interactions between factor X and factor IXa, in addition to the Arg P_1-Asp 189 salt bridge, are two or three hydrogen bonds and several hydrophobic contacts.

The contrast between the two enzyme-substrate complexes described above is quite striking. Whereas the occurrence of two additional specific salt bridges

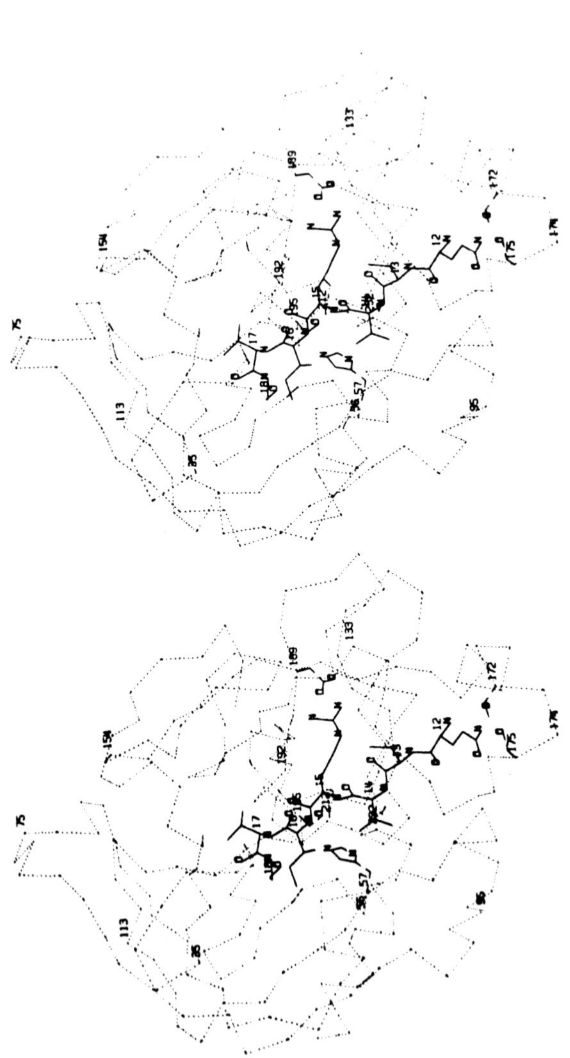

FIGURE 3. Model structure for the factor IXa-factor X complex in stereo. Factor IXa is presented as an α-carbon plot (*dotted lines*) with the relevant side chains (*dashed lines*). The substrate, factor X (*solid lines*), forms the primary specificity salt bridge between Arg P_1 (labeled 15) and Asp 189 of factor IXa. The only other polar interaction lies between Gln P_4 (labeled 12) and Thr 172 and Ser 175. There may be further interactions between the Gln and residues on the adjacent loop around position 99. The side chains of Val P_2 and Val P_3 (labeled 14 and 13, respectively) seem to make very few interactions.

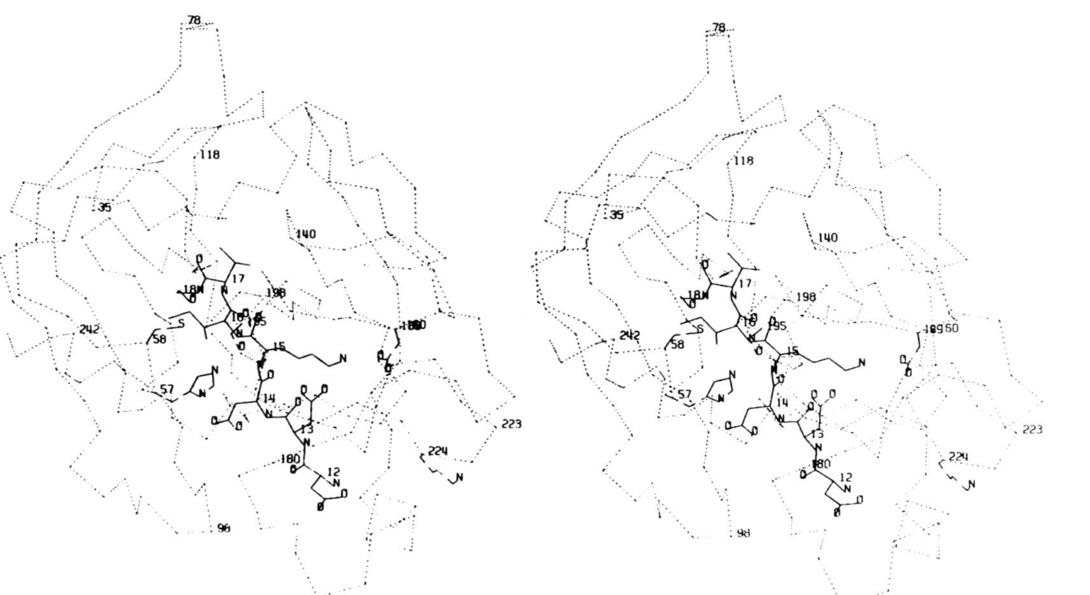

FIGURE 4. Model structure for the complex of the activation peptide of trypsinogen (*solid lines*) with the experimental structure of trypsin (*dotted and dashed lines*) shown in stereo. Lys P_1 (labeled 15) forms a salt bridge with Asp 189 of trypsin. Asp P_2 and P_3 (labeled 14 and 13) point out into solvent as does Asp P_4 (labeled 12). The latter may interact with Lys 224 of trypsin, shown here pointing into solvent also.

beyond the primary specificity interaction seems likely in factor Xa-prothrombin, no such additional charge-charge interactions occur in the factor IXa-factor X complex. In fact, the side chains of Val P_2, Val P_3', and of course, Gly P_3' in factor X seem to contribute very little to either specificity or binding to the factor IXa enzyme.

The very different nature of the interactions found in the factor Xa-prothrombin and factor IXa-factor X complexes encouraged us to examine another enzyme-substrate interaction in order to see how different these substrates are from each other. Therefore, the trypsinogen activation sequence (TABLE 3) was examined in the appropriate conformation for binding to an activating serine protease, in this case taken to be trypsin (see FIG. 4 and TABLE 1). Once again, the primary specificity salt bridge appears, this time between Lys P_1 and Asp 189 of trypsin. The three residues prior to the Lys are aspartates. No countercharge occurs on trypsin for either Asp P_2 or Asp P_3. The ε-amino group of Lys 224 on trypsin is approximately 8 Å from Asp P_4, but can be rotated to lie in close association with that carboxylate. On the other side of Lys P_1, Ile P_1' and Val P_2' lie in their usual, conserved hydrophobic environments. Gly at P_3', of course, has no side chain to interact with the enzyme.

Overall, the trypsinogen peptide seems to interact even less with the trypsin active site region than does factor X with factor IXa. In that latter case, H-bonds were formed to Gln P_4 and valines P_2 and P_3 did provide some hydrophobic interactions beyond the Ile P_1', Val P_2' contacts. In the trypsin-trypsinogen complex, there is almost no interaction of side chains other than that of Ile P_1' and Val P_2'.

This result accords well with the fact that trypsin has lower secondary specificity than the other enzymes studied here. The absence of specific, limiting interactions between the enzyme and substrate side chains at positions P_2 through P_4 and P_3' should allow a wide variety of peptides to bind to the active site. Since trypsin is presumably intended to cleave denatured and partially fragmented proteins, any charged or polar side chains on the substrate which would not interact with the enzyme (such as Asp P_2 through P_4) could be satisfied by the buffer and the solvent.

Comparison of Substrate Properties

It is clear, from examining the three peptide sequences in TABLE 3 and FIGS. 2–4, that each substrate will make quite different hydrophilic and hydrophobic interactions with its respective cleaving enzyme. Consequently,

FIGURE 5. Electrostatic potential surface maps for the three substrate peptides of TABLE 1 shown in stereo. The electrostatic potential is represented as follows: < -15 kcal/mole, red; -15 to -5 kcal/mole, orange; -5 to 5 kcal/mole, green; 5 to 15 kcal/mole, light blue; >15 kcal/mole, blue. *Top* is prothrombin activation peptide, *middle* is factor X activation peptide, and *bottom* is trypsinogen activation peptide. The statistics for these maps are presented in TABLE 4.

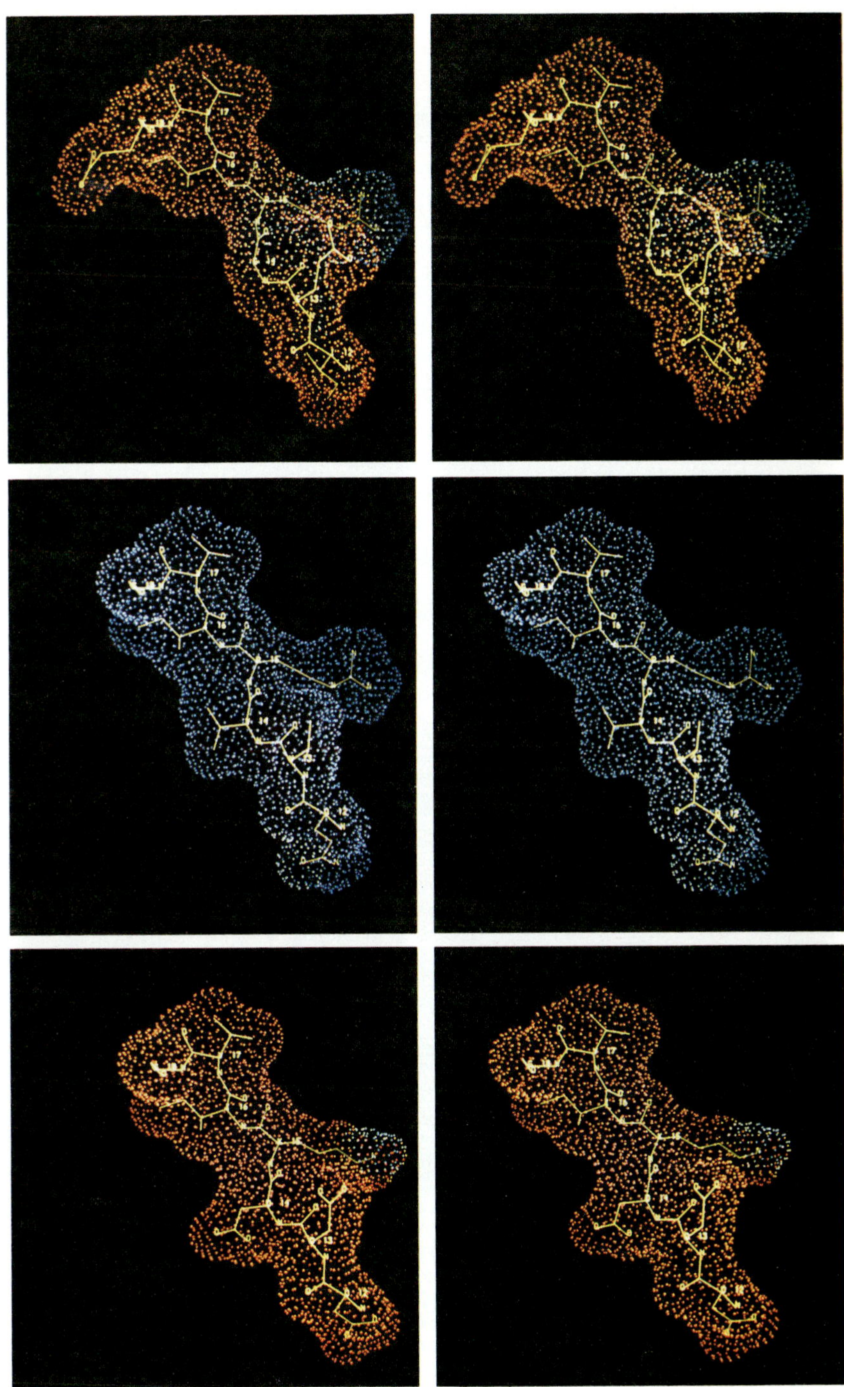

TABLE 4. Distribution of Electrostatic Potential
on the Surface Map of the Substrate

Percent of surface area between	Prothrombin	Factor X	Trypsinogen
< −15 kcal/mole	76.6%	0.0%	95.7%
−15 to −5 kcal/mole	8.1	0.0	1.4
−5 to 5 kcal/mole	5.1	0.0	1.8
5 to 15 kcal/mole	3.1	3.5	1.1
> 15 kcal/mole	7.1	96.5	0.0
Average potential (kcal/mole)	−39.5	41.3	−88.0
Total charge on substrate peptide	−1	+1	−2

it would be useful to compare the different complexes systematically to better understand the nature of their specificity.

Langridge and co-workers[23] have introduced a surface map representation of the electrostatic potential of these molecules as a way of depicting the charge-charge interactions between molecules in a complex. When this method is applied to the substrate peptides (FIG. 5) the differences between the various substrates are highlighted. TABLE 4 gives a breakdown of the per-

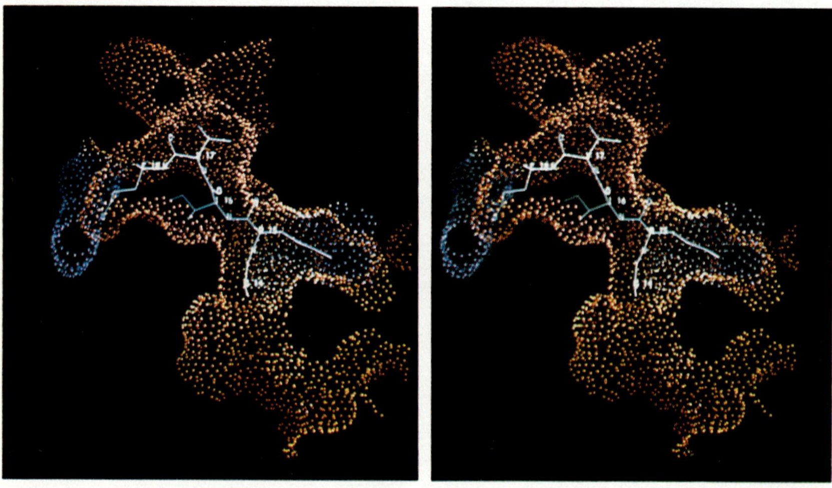

FIGURE 6. A slice through the electrostatic potential surface maps for prothrombin bound to the factor Xa enzyme. The positive surface (*blue*) on the right (labeled R 15) is part of the substrate surface at Arg P_1. It is enveloped by a negative (*red*) surface which is due to the proximity to Asp 189 of factor Xa. On the left, the negative surface is that of Glu P_3' (labeled E 18). Adjacent to it is the positive surface of the enzyme due to Lys 62. Nevertheless, it is very difficult to distinguish which surface comes from the substrate and which from the enzyme, without using different colors for the two surfaces (rather than colors for the electrostatic potential).

cent surface area on each substrate which is positive and negative in order to provide a quantitative indication of the differences between these peptides.

When the electrostatic potential maps are examined in detail, it emerges that the charged side chains, Arg, Lys, Glu, and Asp, dominate the potential function. Therefore, the map for the prothrombin peptide is positive around the Arg P_1 side chain but negative everywhere else because of the two carboxylates of Glu P_3 and P_3' which give the peptide a net charge of -1 (TABLE 4). Similarly, the map for the factor X peptide is almost entirely positive because the only charged species, the positive Arg at P_1, is "felt" everywhere on the surface of the peptide. In the same way, the trypsinogen peptide with a net charge of -2 due to the three negative Asp residues at P_2 to P_4 overwhelms the effect of the positive Lys at P_1 except in the immediate environment of the ε-amino group.

Variation of Substrate-Peptide Electrostatic Potential by Environment

In order to better understand the implications of the electrostatic potential functions, as represented by these color-coded surface maps, to the specificity of the substrate-enzyme interaction, the surface maps of several of the enzymes were examined in the complex with the respective substrate. For this study, the surface maps of the complex of prothrombin with its activating enzyme factor Xa (FIG. 2) are shown in FIG. 6. Unfortunately, the superposition of the two color-coded surfaces makes them difficult to distinguish; consequently, only a thin section can be shown here.[23] It can be seen that, for example, the electrostatic potential on the enzyme surface is opposite to that on the substrate in the region of the salt bridge interactions described above, as shown for Arg P_1-Asp 189 and for Glu P_3'-Lys 62.

Does this mean that the enzyme provides the correct countercharges to balance the charged species on the substrate peptide and yield potential values close to 0? To test this, the electrostatic potential map at the surface of the prothrombin substrate peptide was recomputed, but with the three countercharge residues, Asp 189, Arg 143, and Lys 62 included to give a total charge of zero for the substrate + countercharges system. The resulting map is shown in FIG. 7B and summarized in TABLE 5. The charge distribution is now strikingly different than for the substrate peptide by itself (FIG. 7A). The first point to stress is that very little of the surface area, 13%, is close to 0 potential. Thus, the countercharges are not neutralizing the charge on the substrate but are changing the electrostatic potential distribution considerably. The surface about P_1' to P_3' changes from negative to positive because of its proximity to Lys 62, Arg P_1, and Arg 143. Similarly, the tip of the surface for Arg P_1 becomes negative because of the close approach of Asp 189. Some of the molecular surface does reflect the nature of the atoms that are immediately nearby, rather than longer range charge effects, as for example, the side chain of Ile P_4 is close to zero whereas its carbonyl oxygen is negative.

Having seen the effect of including just the countercharges, how is the electrostatic potential of substrate peptide affected by being placed in the

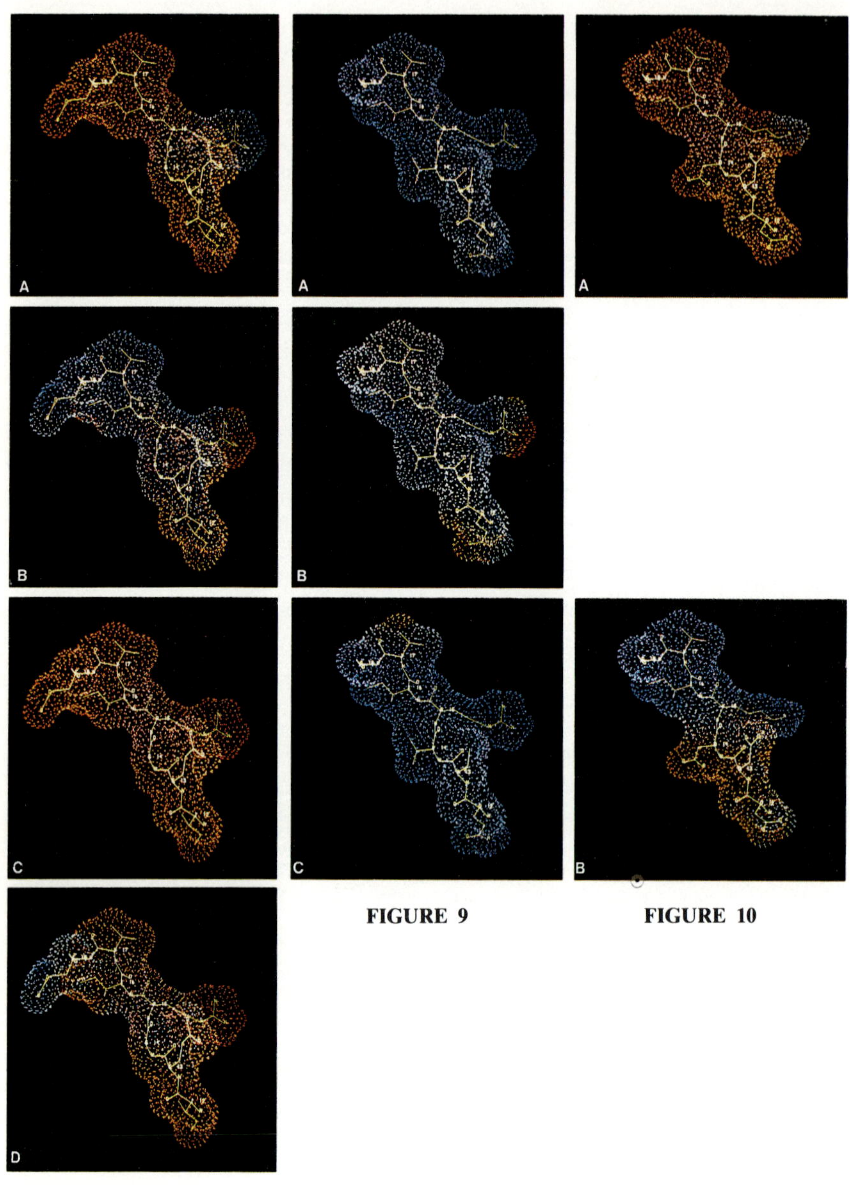

FIGURE 9

FIGURE 10

FIGURE 7

TABLE 5. Distribution of Electrostatic Potential
for the Prothrombin Substrate Peptide

Percent of surface area between	Prothrombin	Prothrombin + 3 countercharges	Prothrombin + factor Xa	Prothrombin + factor Xa + Ca^{++}
< -15 kcal/mole	76.6%	24.4%	99.9%	70.1%
-15 to -5 kcal/mole	8.1	11.1	0.1	13.3
-5 to 5 kcal/mole	5.1	13.1	0.0	8.2
5 to 15 kcal/mole	3.1	16.5	0.0	6.5
> 15 kcal/mole	7.1	34.9	0.0	1.9
Average potential (kcal/mole)	-39.5	2.6	-58.4	-27.3
Net charge on system	-1	0	-1	$+1$

complete environment of the factor Xa enzyme model structure? The net charge of the factor Xa molecule is 0. This gives a total charge of -1 for the enzyme-substrate complex. The result of this calculation was a complete surprise; FIG. 7C and TABLE 5 present the results. The entire map is negative with an average electrostatic potential of -58 kcal/mole. Thus, the peptide in the active site of the enzyme must lie in a highly negative environment

---

FIGURE 7. Electrostatic potential surface maps for the substrate prothrombin activation peptide (see TABLE 5). **(A)** Potential map for prothrombin peptide by itself with a net charge of -1. **(B)** Map for the prothrombin peptide with the three salt bridge residues: Lys 62, Arg 143, and Asp 189. The net charge on this system is 0. **(C)** Potential surface map for the prothrombin peptide on the factor Xa enzyme. Note that the whole surface is now highly negative with mean potential value of -58 kcal/mole. This is due to the distribution of charged residues on the enzyme (see TABLE 6 and text for discussion). **(D)** Ca^{++} placed 14 Å from the substrate near residues Glu 36 and Glu 37 of factor Xa (see FIG. 8). Note that the positive surface corresponds exactly to the carboxylate side chain of Glu P_3' (labeled E 18) which is itself negative.

FIGURE 9. Electrostatic potential surface maps for the substrate factor X activation peptide (see TABLE 7). **(A)** The map for factor X peptide which has a net charge of $+1$. The map is dominated by this positive charge of Arg P_1. **(B)** Potential map for the peptide together with Asp 189, the countercharge for Arg P_1 (labeled R 15). The net charge is now 0. A discussion of this distribution appears in the text. **(C)** Map for the factor X peptide together with the model for its activating enzyme, factor IXa. In the factor IXa enzyme, the active site and specificity sites are highly positive, in contrast to factor Xa (FIG.7C), where the environment was highly negative. The potential at the surface of residue Val P_2' (labeled V 17) is negative, due to a concentration of negatively charged residues in the region of the enzyme adjacent to this position (see text).

FIGURE 10. Electrostatic potential surface maps for the substrate trypsinogen activation peptide (see TABLE 8). **(A)** The map for the trypsinogen peptide alone with a charge of -2 due to the Asp residues. The surface is overwhelmingly negative. **(B)** The trypsinogen peptide in the environment of trypsin gives this potential map. The *N*-terminal portion remains negative since there are no countercharges for Asp P_2 and Asp P_3 (labeled D 14 and D 13, respectively) and they point out into solvent. There is a countercharge for Asp P_4 (labeled D 12) which is Lys 224, and thus some positive surface appears at this residue. The *C*-terminal portion is apparently dominated by the large positive net charge on the trypsin molecule of $+10$. Note that the transition from negative to positive occurs right at the point of the cleaved peptide bond.

TABLE 6. Net Charge Distribution on Enzyme in Shells about the Substrate

Substrate	Prothrombin		Factor X		Trypsinogen	
Total charge	-1		$+1$		-2	
Enzyme	Factor Xa		Factor IXa		Trypsin	
Total charge	0		$+1$		$+10$	
Shell range (Å)	Net charge in shell	Accumulated net charge	Net charge in shell	Accumulated net charge	Net charge in shell	Accumulated net charge
0–3	0	0	-1	-1	0	0
3–5	-1	-1	0	-1	-2	-2
5–7	-1	-2	-1	-2	$+1$	-1
7–9	0	-2	$+3$	$+1$	0	-1
9–12	0	-2	$+2$	$+3$	$+4$	$+3$
12–15	$+1$	-1	$+1$	$+4$	$+3$	$+6$
15–20	$+2$	$+1$	$+1$	$+5$	$+1$	$+7$
20–50	-1	0	-4	$+1$	$+3$	$+10$

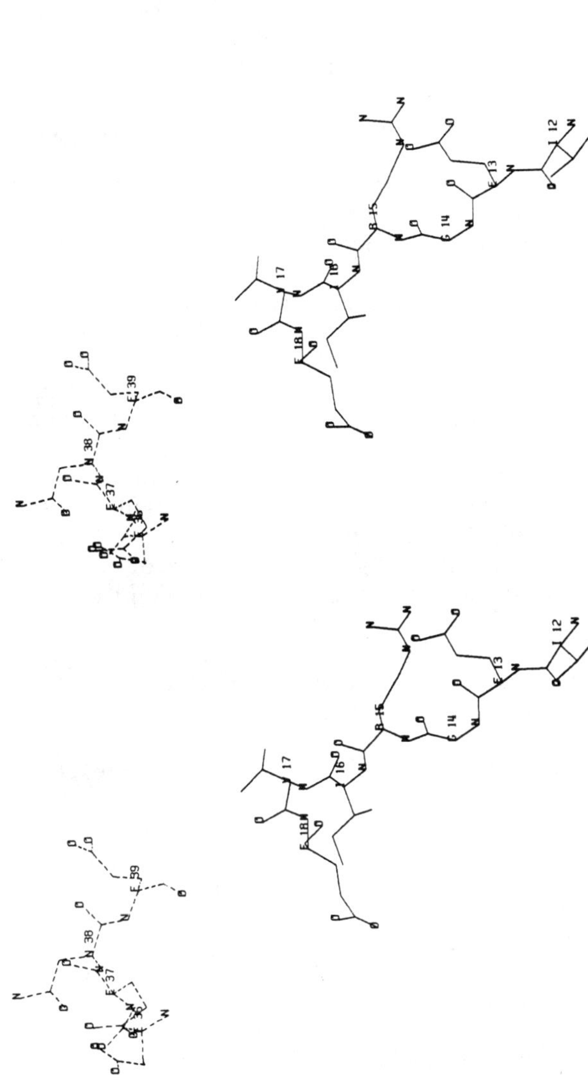

FIGURE 8. The loop of factor Xa containing residues Glu 36, Glu 37 and Glu 39 is shown (*dashed lines*) in relation to the prothrombin substrate (*solid lines*). The position of the added Ca^{++} ion is marked by a "#" in upper left of the figure near the carboxylates of Glu 36 and Glu 37. The nearest distance from the Ca^{++} ion to Glu P_3' (labeled E 18) of prothrombin is 14.4 Å.

even though, as noted above, the net charge on the enzyme is 0. In order to determine why this is the case, the net charge on the enzyme was calculated in shells around the substrate, as shown in TABLE 6. As can be seen, factor Xa has an excess of negative charges within 7 Å of the substrate. There are shells with an excess of positive charge also, but these lie much farther away, from 12 to 20 Å, and thus by Coulomb's law, have a much smaller effect on the substrate. It is particularly striking that the active-site environment of factor Xa is so negative, when the net charge on the prothrombin peptide is also negative.

Is this highly negative environment of the active site a realistic view of this enzyme-substrate interaction in the factor Xa-prothrombin complex? Clearly, these charge effects could be modified and modulated by the binding of ions from solvent and buffer onto the protein. In particular, there is a loop in factor Xa that contains three Glu residues very close together, at positions 36, 37 and 39 (FIG. 8). Two, and perhaps all three, of these residues may form a Ca^{++} ion binding site. In order to test the effect of the binding of such an ion, a Ca^{++} was bound to the enzyme at this site and given a charge of $+2.0$. It is important to point out that the closest this ion approaches to the substrate peptide is 14.1 Å at Glu $P_3{}'$ (see FIG. 8).

The electrostatic potential map of the prothrombin activation peptide was recalculated in the environment of the factor Xa enzyme model with the one Ca^{++} ion bound, as described. The resulting map is shown in FIG. 7D and is summarized in TABLE 5. There is a significant influence of this single Ca^{++} ion on the electrostatic potential in the environment of Glu $P_3{}'$ and its surrounding, which has changed from significantly negative to positive. Thus, it is clear that a proper treatment of the electrostatic environment of the substrate must take into account solvent and buffer ion interaction with the enzyme even at sites somewhat removed from the immediate region of the active and specificity sites of the enzyme.

Similar calculations were performed on the electrostatic potential map of the factor X activation peptide substrate bound to its activating enzyme factor IXa (see FIG. 9 and TABLE 7). As previously discussed, the substrate peptide appears to be entirely positive with a mean electrostatic potential value of 41 kcal/mole (Fig. 9A) due to the Arg at the P_1 site. When the countercharge for this Arg, Asp 189 from the protein, is included in the calculation of the electrostatic potential, the result is a significant difference in the potential function (Fig. 9B and TABLE 7) which now has an average potential value of 8.4 kcal/mole. Very little of the surface is actually negative, mostly in the immediate region of Asp 189. There is a positive band that is largely due to the influence of the Arg P_1 guanidinium group. Moving further away from Arg P_1, there is a belt of approximately neutral or slightly positive or negative potential that is largely the result of the carbonyl oxygens of residues P_2 and P_4 and the Gln P_4 side chain. Finally the edge of the molecule which points mostly away from the activating enzyme (not easily seen in this view) is again positive. This appears to be due to a series of main chain amino hydrogens of residues P_2, P_1 and $P_1{}'$ all pointing in a similar direction (see FIG. 3). Once again the addition of a countercharge, which gives a formally

TABLE 7. Distribution of Electrostatic Potential
for the Factor X Substrate Peptide

Percent surface area between	Factor X	Factor X + 1 countercharge	Factor X + factor IXa
< -15 kcal/mole	0.0%	4.3%	0.0%
-15 to -5 kcal/mole	0.0	7.7	1.9
-5 to 5 kcal/mole	0.0	24.7	1.9
5 to 15 kcal/mole	3.5	36.1	3.0
> 15 kcal/mole	96.5	27.3	93.2
Average potential (kcal/mole)	41.3	8.4	64.9
Net charge on system	1	0	2

net uncharged system, results in a distribution of very significant positive and negative potentials rather than a close to zero value.

Next, the electrostatic potential map was recalculated for the factor X peptide in the environment of the factor IXa model structure (FIG. 9C and TABLE 7). The resulting map is different from both the isolated peptide and the peptide-countercharge distribution. The potential map is almost entirely positive. This is not the result of the overall $+1$ charge on the factor IXa molecule (versus -1 for factor Xa, see TABLE 6) because the addition of $+2$ to factor Xa (FIG. 7D) had only a limited effect on the negative environment even though the total net charge of factor Xa $+$ Ca^{++} is $+1$. Rather, it is once again the result of the charge distribution on factor IXa (TABLE 6). Whereas the immediate region about the substrate from 0–7 Å tends to be negative, a larger net positive charge from 7–15 Å seems to overwhelm the negative group and gives the active site a positive character.

Interestingly, despite the overall positive environment, the surface about the side chain and carbonyl oxygen of Val P$_2'$ is quite negative. This is due to a concentration of negative charges and a paucity of positive charges in this region of the factor IXa enzyme, in particular, Glu 36 and Glu 74 but also Glu residues at positions 70, 78, and 80.

TABLE 8. Distribution of Electrostatic Potential
for the Trypsinogen Substrate Peptide

Percent surface area between	Trypsinogen	Trypsinogen + trypsin
< -15 kcal/mole	95.7%	30.1%
-15 to -5 kcal/mole	1.4	9.4
-5 to 5 kcal/mole	1.8	5.9
5 to 15 kcal/mole	1.1	3.4
> 15 kcal/mole	0.0	51.2
Average potential (kcal/mole)	-88.0	21.3
Net charge on system	-2	$+8$

The average potential of 65 kcal/mole for factor X on the factor IXa enzyme indicates that the factor X substrate peptide, which itself has a net positive charge, resides in a highly positive environment in factor IXa. This is the exact parallel of the situation in the enzyme factor Xa, which was a negative substrate peptide in a highly negative environment (see Fig. 7C). Thus, in each case, the enzyme active site carries a potential that has the same sign as the potential on the substrate peptide. One might speculate that this may represent a mechanism whereby the enzyme prevents too strong binding of the substrate—which would result in diminished turnover.

Similarly dramatic differences can be observed in the electrostatic potential function of the trypsinogen activation peptide on the enzyme trypsin (Fig. 10 and Table 8). In the isolated substrate peptide, as noted above, the three Asp residues at P_2 through P_4 dominate the potential function, virtually overwhelming the positive charge at Lys P_1. Because countercharges for Asp P_2, Asp P_3 and Asp P_4 (except perhaps Lys 224 for the latter, see Fig. 4), do not occur on trypsin, the potential map for the substrate peptide together with enzyme countercharges cannot be computed. However, the addition of the enzyme which has a net $+10$ charge has a remarkable effect on the potential function. The N-terminal half of the substrate peptide remains negative, due presumably to the aspartate residues. Only Asp P_4 has a possible countercharge on the enzyme in residue Lys 224, hence the positive and zero region at the surface of residue P_4. The C-terminal half of the molecule does apparently "feel" the highly positive nature of the molecule. The complete substrate is not dominated by the $+10$ charge of trypsin because most of the positive side chains of trypsin are distant from the substrate (see Table 6). It is especially clear that both with respect to the Asp residues at positions P_2 through P_4 and to the many excess positive residues on trypsin, all of which are exposed to solvent, the binding of ions from the buffer and solvent should play a crucial role in the proper evaluation of the electrostatics of substrate-enzyme interaction.

Because trypsin is a digestive enzyme, it tends to be less specific than the other serine proteases studied here. Therefore, it probably operates mostly on denatured proteins or highly accessible peptide portions of macromolecules. Thus, it is likely that a substrate peptide such as in trypsinogen will bind to trypsin with Asp residues P_2 and P_3 and perhaps even P_4 (despite the Lys 224 counterion) interacting ultimately with solvent and buffer ions. Hence, the electrostatic effect of these aspartates would be considerably diminished. It will be interesting to examine several other possible substrates of trypsin in order to see how they are affected by the trypsin environment and compare these results with those obtained for the more specific blood clotting factors described above.

ACKNOWLEDGMENTS

I thank Drs. T. J. O'Donnell and Arthur Olson for providing the GRAMPS graphics program and Dr. Michael Connolly for the solvent-exclusion surface program. This study also benefited from graphics programs written by Fred Thomas and Jeff Wakat.

REFERENCES

1. BROWNE, W. J., A. C. T. NORTH, D. C. PHILLIPS, K. BREW, T. C. VANAMAN & R. L. HILL. 1969. J. Mol. Biol. **42**: 65–86.
2. MCLACHLAN, A. D. & D. M. SHOTTON. 1971. Nature New Biol. **229**: 202–205.
3. JURASEK, L., R. W. OLAFSON, P. JOHNSON & L. B. SMILLIE. 1976. *In* Proteolysis and Physiological Regulation. D. W. Ribbons & K. Brew, Eds.: 93–123. Academic Press. New York.
4. KRETSINGER, R. H. 1976. Annu. Rev. Biochem. **45**: 239–266.
5. GREER, J. 1980. Proc. Nat. Acad. Sci. U.S.A. **77**: 3393–3397.
6. GREER, J. 1981a. J. Mol. Biol. **153**: 1027–1042.
7. GREER, J. 1981b. J. Mol. Biol. **153**: 1043–1053.
8. BLUNDELL, T., B. L. SIBANDA & L. PEARL. 1983. Nature (London) **304**: 273–275.
9. BIRKTOFT, J. J. & D. M. BLOW. 1972. J. Mol. Biol. **68**: 187–240.
10. TULINSKY, A., N. V. MANI, C. N. MORIMOTO & R. L. VANDLEN. 1973. Acta Crystallogr. **29**(B): 1309–1322.
11. HUBER, R., D. KUKLA, W. BODE, P. SCHWAGER, K. BARTEL, J. DIESENHOFER & W. STEIGEMANN. 1974. J. Mol. Biol. **89**: 73–101.
12. STROUD, R. M., L. M. KAY & R. E. DICKERSON. 1974. J. Mol. Biol. **83**: 185–208.
13. SHOTTON, D. M. & H. C. WATSON. 1970. Nature (London) **225**: 811–816.
14. SAWYER, L., D. M. SHOTTON, J. W. CAMPBELL, P. L. WENDEL, H. MUIRHEAD, H. C. WATSON, R. DIAMOND & R. C. LADNER. 1978. J. Mol. Biol. **118**: 137–208.
15. TITANI, K., K. FUJIKAWA, D. L. ENFIELD, L. H. ERICSSON, K. A. WALSH & H. NEURATH. 1975. Proc. Nat. Acad. Sci. U.S.A. **72**: 3082–3086.
16. KATAYAMA, K., L. H. ERICSSON, D. L. ENFIELD, K. A. WALSH, H. NEURATH, E. W. DAVIE & K. TITANI. 1979. Proc. Nat. Acad. Sci. U.S.A. **76**: 4990–4994.
17. SCHECHTER, I. & A. BERGER. 1967. Biochem. Biophys. Res. Commun. **27**: 157–162.
18. BLOW, D. M., J. JANIN & R. M. SWEET. 1974. Nature (London) **249**: 54–57.
19. SWEET, R. M., H. T. WRIGHT, J. JANIN, C. H. CHOTHIA & D. M. BLOW. 1974. Biochemistry **13**: 4212–4228.
20. SEGAL, D. M., G. H. COHEN, D. R. DAVIES, J. C. POWERS & P. E. WILCOX. 1971a. Cold Spring Harbor Symp. Quant. Biol. **36**: 85–90.
21. SEGAL, D. M., J. C. POWERS, G. H. COHEN, D. R. DAVIES & P. E. WILCOX. 1971b. Biochemistry **10**: 3728–3738.
22. JAMES, M. N. G., G. D. BRAYER, L. T. J. DELBAERE, A. R. SIELECKI & A. GERTLER. 1980. J. Mol. Biol. **139**: 423–438.
23. LANGRIDGE, R., T. E. FERRIN, I. D. KUNTZ & M. L. CONNOLLY. 1980. Science **211**: 661–666.
24. CONNOLLY, M. L. 1983. Science **221**: 709–713.
25. BLANEY, J. M., P. K. WEINER, A. DEARING, P. A. KOLLMAN, E. C. JORGENSEN, S. J. OATLEY, J. M. BURRIDGE & C. C. F. BLAKE. 1982. J. Am. Chem. Soc. **104**: 6424–6434.
26. KOLLMAN, P., P. WEINER & A. DEARING. 1981. Ann. N.Y. Acad. Sci. **367**: 250–268.
27. GELIN, B. & M. KARPLUS. 1979. Biochemistry **18**: 1256–1268.
28. WARSHEL, A. & M. LEVITT. 1976. J. Mol. Biol. **103**: 227–249.
29. HARTLEY, B. S. 1970. Phil. Trans. Roy. Soc. (series B) **257**: 77–87.

Generation of Nucleic Acid Structures and Binding of Molecules to DNA[a]

K. J. MILLER, FREDERICK H. REIN, ERIC R. TAYLOR, AND PAUL J. KOWALCZYK

Department of Chemistry
Rensselaer Polytechnic Institute
Troy, New York 12180

INTRODUCTION

The analysis of molecules that exhibit specific biological functions such as antitumor or carcinogenic activity requires knowledge of the mode of action on a molecular level. Biochemical processes depend on the fundamental building blocks and functional groups of these molecules. The importance of molecular structure has been emphasized in the chemists' search for the so-called structure-activity relationships. However, the relationship between a chemical structure and the biochemical activity also depends on the receptor sites to which these molecules bind. Therefore, a modern concept of structure must include the consideration of both a donor molecule and the receptor site. The structural details of the latter are often difficult to define. Fortunately, it is known that the DNA is the receptor site for many antitumor and carcinogenic agents, and this fact is used in our theoretical studies of donor-receptor or drug-DNA interactions. The study of molecules exhibiting activity with a receptor site enables one to identify the important structural ingredients which may be responsible for activity. Our research is based on two generally accepted ideas. First, the receptor site DNA is at the center of the activity being studied. Second, certain intercalating drugs are a group of molecules which are antitumor agents, and some polynuclear aromatic hydrocarbons are metabolized and then covalently bind to DNA.

The process of intercalation is characterized by the insertion of a planar molecule between adjacent base pairs of the DNA.[1,2] Because many of these molecules exhibit antitumor activity and also inhibit replication of the DNA, it is believed that intercalation is an important step in arresting cancer through chemotherapy. These concepts have been recently reviewed.[3] The diol epoxide of benzo[a]pyrene (BPDE) is known to intercalate.[4] However, it reacts mainly with N2 on guanine to form a covalent bond.[5-21] This is believed to be the

[a] This work was supported by PHS grant no. CA-28924 awarded by the National Cancer Institute, DHHS, and by a grant of computer time from Rensselaer Polytechnic Institute.

carcinogenic event. Modeling these interactions requires methods to alter the DNA, to adjust the molecule in the binding site, and to visualize the process.

Several developments in our research will be discussed. First, results of a generalization of our computer-assisted algorithm to generate nucleic acid structures (AGNAS)[22] will be presented. The AGNAS technique determines which backbone conformations can be constructed for a given orientation of bases and/or sugar units with given puckers. This is demonstrated by the conformations obtained when base pairs are stacked at 3.38 Å to yield B-DNA, when base pairs are kinked,[23] when localized regions of the backbone are rotated via crankshaft motion and when bases are rotated about their glycosidic bonds. Second, the optimum binding orientations of intercalation complexes and covalently bound adducts are illustrated for donor-receptor complexes. For the intercalation process a sequence of substituted anthraquinones will be used to demonstrate the use of symmetry in the analysis of binding and in drug design. Alterations of the local conformation such as kink in B-DNA will be demonstrated to model the BPDE adduct with N2 on guanine. Third, examples of computer graphics techniques will be presented throughout the paper, and then discussed.[24-26] The binding site will be illustrated with a steric surface on which electrostatic contours are drawn to indicate the relatively positive and negative potential regions.

The binding of molecules to DNA is studied by calculating the minimum energy for one of the altered DNA structures. The energy for the binding process

$$\text{B-DNA} \xrightarrow{\Delta E^{\nu}_{\text{DNA}}} \text{Receptor Site } \nu \xrightarrow{\Delta E^{\nu}_{\text{MOLC}}} (\text{Site } \nu)\bullet\text{MOLC}$$

is partitioned into contributions for the formation of receptor site ν in DNA, and for the binding of a molecule, MOLC, to DNA

$$\Delta\varepsilon^{\nu}_{\text{MOLC}} = \Delta E^{\nu}_{\text{DNA}} + \Delta E^{\nu}_{\text{MOLC}} \tag{1}$$

where $\Delta E^{\nu}_{\text{DNA}}$ is the energy required to open B-DNA to a receptor site and

$$\Delta E^{\nu}_{\text{MOLC}} = I^{\nu}_{\text{MOLC}} + \Delta C^{\nu}_{\text{MOLC}} \tag{2}$$

where I^{ν}_{MOLC} is the interaction energy between the DNA and MOLC, and $\Delta C^{\nu}_{\text{MOLC}}$ is the change in conformational energy required to alter the molecule to fit into site ν. For a particular calculation a representative receptor site corresponding to the minimum energy conformation in each of the three intercalation sites or in one of the kinked sites is used. For each site, the molecule is adjusted until a minimum for $\Delta\varepsilon^{\nu}_{\text{MOLC}}$ is achieved.

GENERATION OF NUCLEIC ACID STRUCTURES

Modeling of nucleic acid structures by various mathematical methods[22,27-30] has been extensively employed in the study of DNA flexibility and interactions with various small molecules. The algorithm developed in this labora-

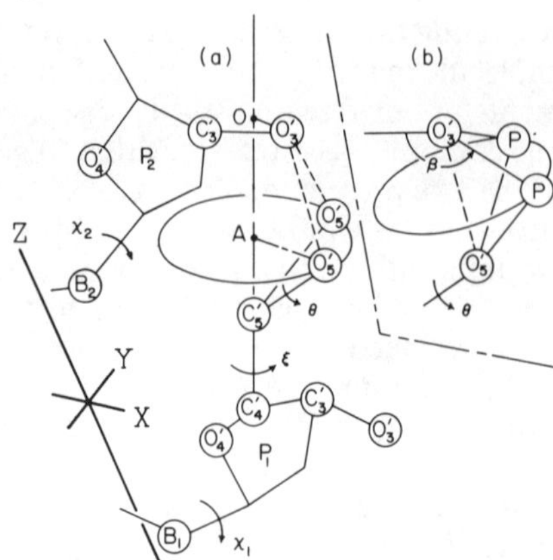

FIGURE 1. AGNAS technique. For a specified orientation of bases B_1 and B_2, the glycosidic angles χ_1 and χ_2 orient the sugars with puckers p_1 and p_2. O_5' and P must be placed in the DNA. **(a)** Two positions of O_5' in its circular orbit about the $C_4'C_5'$ axis for which the distance $O_5'O_3'$ is the experimental value. **(b)** Two positions of P in its circular orbit about one of the $O_5'O_3'$ axes for which $(C_5'O_5',P)$ is the experimental value.

tory to generate nucleic acid structures relies on a vector approach for the construction of a backbone after the bases in adjacent base pairs have been oriented. The resultant structures contain experimentally determined bond angles and bond lengths, and must satisfy van der Waals criteria for contacts between nonbonded atoms.

The AGNAS technique provides a method to determine whether a backbone can be constructed after the bases and sugars on adjacent base pairs have been oriented. Therefore a set of independent parameters must be specified. They are chosen to perform the following operations: (1) orient bases within a base pair via a twist, tilt and bulge of individual bases, (2) orient two adjacent base pairs via a displacement and rotation, (3) specify the sugar puckers and glycosidic angles. In FIG. 1 the procedure is illustrated. Steps 1–3 and the geometrical requirements imposed by experimental bond lengths, bond angles and atomic hybridization can orient all but the O5' and P atoms in space. These two atoms must be placed in the backbone with acceptable bond lengths and angles through these atoms to yield an acceptable conformation and a possible receptor site. First O5' is rotated about an axis through (C4',C5'). At most two positions of O5' on the circle yield an acceptable (O5',O3') distance. For each position of O5', P is rotated about an axis through (O5',O3'). At most two positions on this circle yield the required (O5',P,O3') angle. Because the backbone has been constructed toward O3', the (P,O3',C3') angle has not been specified yet. For each of the geometrical structures produced by this sequence of steps, the angle (P,O3',C3') must be examined. If it does not correspond to the experimental value, then the conformation is rejected. By systematically changing certain parameters, an acceptable conformation can be constructed when (P,O3',C3') = 119.4° is achieved. Several examples follow. If the base pairs are assumed to be planar, separated by 3.38 Å and rotated by 36°, the B-DNA structure can be reproduced by sweeping through all glycosidic angles.[22] Three geometrical structures are obtained, but only one has acceptable nonbonded contacts. If adjacent base pairs are assumed to be planar, separated by 6.76 Å and required to fit into B-DNA in a tetramer duplex unit, then three families of intercalation sites can be obtained by sweeping through all helical angles. If adjacent base pairs are kinked, then physically meaningful conformations can be obtained by sweeping through the helical angle. Variations of the B-DNA as well as other conformations can be accomplished in several ways without using the AGNAS technique. First, an *anti→syn* transformation of individual bases can be achieved by a rotation about the glycosidic bond. Second, the occurrence of collinear bonds along the backbone can allow a rotation of all atoms located between these bonds in a crankshaft motion.

After a single strand of the DNA has been constructed, advantage is taken of the antiparallel nature of the DNA. The complementary strand is generated by assuming pseudodyad symmetry. Although convenient, this assumption is not necessary as will be illustrated by the *anti→syn* rotation of a base and crankshaft transformation. In general, it is not easy to break the dyad symmetry while maintaining Watson-Crick base pairing. Results of applications of the AGNAS technique are presented in this paper.

COVALENT INTERCALATIVE BINDING

The carcinogenicity of polynuclear aromatic hydrocarbons (PAHs) involves their conversion to active metabolites[31] and covalently bound adducts[5-18] with DNA. The arene oxides are among the most reactive,[32-34] and it is assumed that their reaction with cellular macromolecules is a critical step in the carcinogenic event. Of the four diastereoisomers of the diol epoxides of benzo[a]pyrene, (+)7β,8α-dihydroxy-9α,10α-epoxy-7,8,9,10-tetrahydrobenzo[a]pyrene, BPDE I(+), is carcinogenic. This molecule has received much attention.[5-21] It intercalates[4] and covalently binds to N2 on guanine.[5-21] To understand the structural characteristics of the receptor site, the AGNAS technique was utilized to search for conformations of DNA that yielded complexes consistent with experimental data. Geacintov and co-workers[4] find that the diol epoxide is intercalated, that a tetrol forms, and that there is no covalent bond formation with N2 on guanine while the pyrene moiety is intercalated. However, after adduct formation with N2 on guanine, they find that the pyrene is oriented at 35° relative to the helical axis[35] (*i.e.*, not perpendicular as found in classical intercalation). Hogan *et al.*[36] find that the pyrene moiety is also not perpendicular to the helical axis, but also observe that the DNA is bent at the site of adduct formation. Gamper *et al.*[37] observe that the DNA is unwound after adduct formation. In the present investigation, a series of receptor sites are presented to rationalize these experimental results.

Proposals[38,39] to place a covalently bound BPDE I(+) adduct in the minor

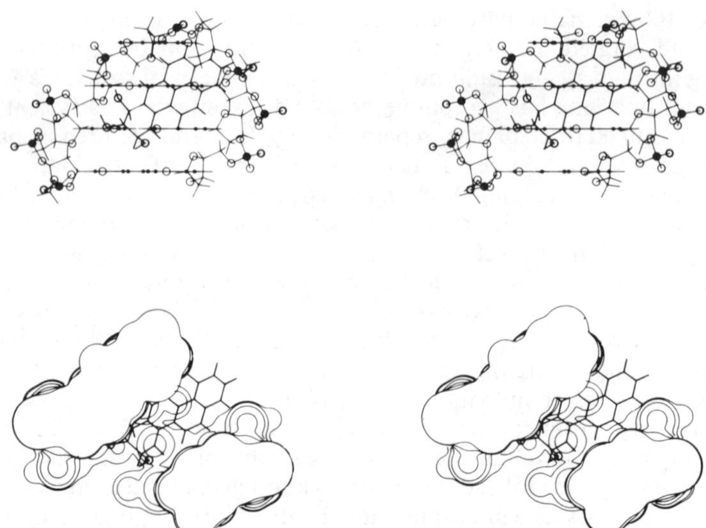

FIGURE 2. Illustration of steric interference by omission of a portion of the chemical structure when BPDE I(+) is assumed to bind to N2 on guanine of DNA in the B-conformation. *Trans* addition is shown. The view is along an axis perpendicular to the helical axis of the DNA and an axis containing the glycosidic carbon atoms C_1' of the second base pair from the bottom. Oxygen (O), phosphorous (●) and nitrogen (•) are denoted.

groove of B-DNA lead to poor nonbonded contacts[24,40] as illustrated in FIG. 2. Although the adduct has been adjusted to yield an optimum fit, the pyrene moiety penetrates the region of the DNA represented by the (space-filling) steric contours. This demonstration of a poor fit is a feature of hidden line removal in the graphics techniques developed in our laboratory. A search for conformations with increased space to accommodate the pyrene moiety led to unwound DNA of the type used for intercalation, and families of kinked DNA.[23] The latter is particularly interesting because a kinked structure with base pairs separated by 7.6 Å can accommodate the adduct between the base pairs in a covalent intercalative mode. In FIGS. 3 and 4, a sequence of BPDE I(+) complexes with DNA is shown. Classical intercalation of the BPDE I(+) yields an optimum orientation with the reactive epoxide adjacent to the N2 on guanine. Trigonal and/or tetrahedral hybridization cannot be achieved in this receptor site. A further separation and kinking of base pairs yields a receptor site in which the pyrene moiety is approximately parallel to one base pair and the covalent adduct occurs with proper hybridization about the C10 (on BPDE) and N2 (on guanine) atoms. We have reported structures with kinks ranging from 0° to 70° for opening into the major groove.[23] Two optimized orientations of the adduct are possible, (5′) and (3′), depending on the strand to which guanine is bound (FIGS. 4a and 4b, respectively).

From the theoretical results of Miller *et al.*[40] and experimental studies

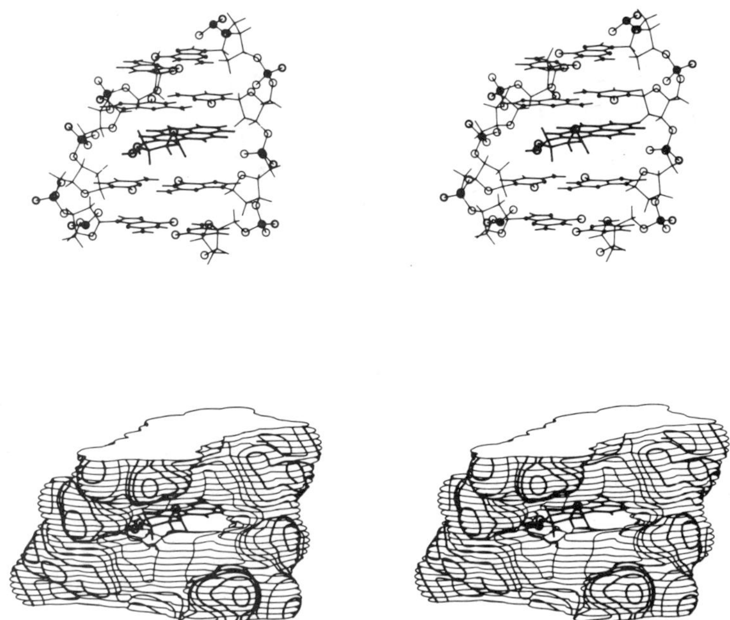

FIGURE 3. BPDE I(+) in its optimum orientation for classical intercalation.

of Geacintov and co-workers[35] concerning the externally bound adduct, a closer examination of the problem revealed that the B-DNA can accommodate the BPDE I(+) covalent adduct to N2 on guanine if the host guanine undergoes an *anti→syn* transition in its glycosyl torsion. A minimum energy conformation with $\chi_{syn} = \chi_{B\text{-}DNA} + 200°$ is observed. The N2 of guanine is displaced to the outside of the DNA helix and can, in this orientation, accommodate the bound BPDE I(+) adduct as illustrated in Fig. 5. Any *syn* orientation of the guanine less than $\chi_{B\text{-}DNA} + 200°$ results in severe nonbonded crowding of the guanine and BPDE I(+) adduct atoms with the sugar phosphate backbone.

Such a structure is well within 10 kcal/tetramer mole in energy of the kinked forms illustrated in Figs. 4a and 4b. Though the models and experimental data tend to favor an internally bound BPDE I(+) adduct at a site that disrupts the regular alignment of the helical axes on either side of the binding site, an externally bound BPDE I(+) adduct does have merit, particularly where a BPDE I(+) adduct bound to a duplex of slightly altered B-DNA is carried along in replication or transcription. During intermediate changes, such as a locally denatured duplex, the constraints of Watson-Crick base pairing are disrupted, leaving the host guanine in an orientation that preserves the overall B-DNA structure.

Another transformation that locally modifies one strand of the DNA results from crankshaft motion, *i.e.,* a rotation of atoms about collinear

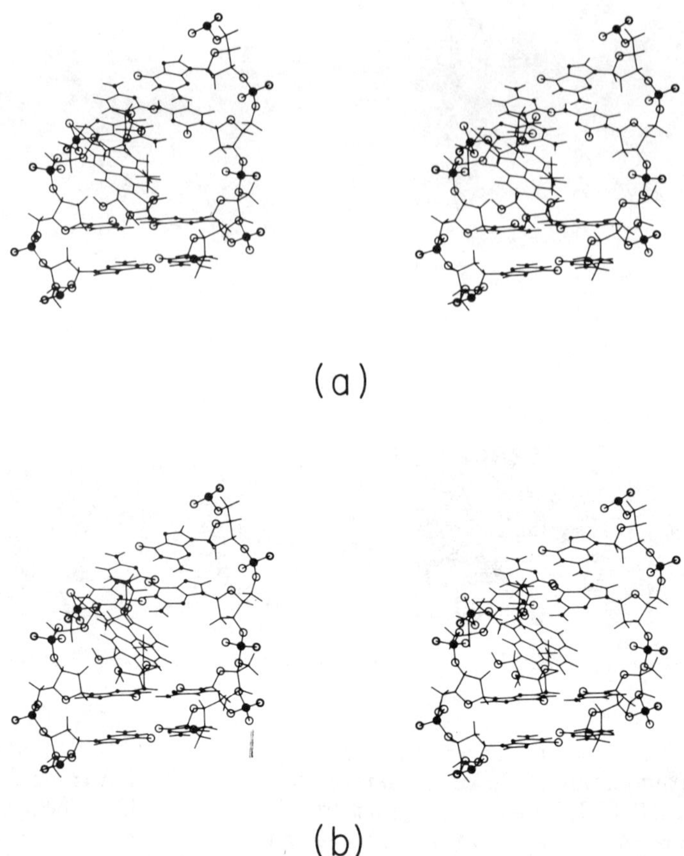

(a)

(b)

FIGURE 4. Covalent intercalative-bound BPDE adducts in DNA with C(2')-*endo*-C(3')-*endo* sugar puckers. (a) 5' bound adduct on a ↓G•C↓ base pair and (b) 3' bound adduct on ↑G•C↓ base pair. In both structures, the kink site bends the tetramer duplex at an angle of 39°. Space-fixed coordinate axes are shown.

bonds. For example, in B-DNA, the O3'-C3' and O5'-P bonds of a nucleotide are sufficiently collinear to permit the O3' and P atoms to serve as end points of a rotation axis. Rotation about such an axis produces an extrahelically displaced base and sugar. Bending of the P-O3'-C3' and O5'-P-O3' bond angles at each end of the crankshaft axis is less than 9° on the average, and such a distortion can be distributed over the entire backbone. In FIG. 6, a trimer duplex is illustrated for normal B-DNA and after local modification by a crankshaft motion of 180°.

INTERCALATION

The antitumor activity of molecules possessing a planar chromophore involves a mechanism in which intercalation appears to play a role. Although

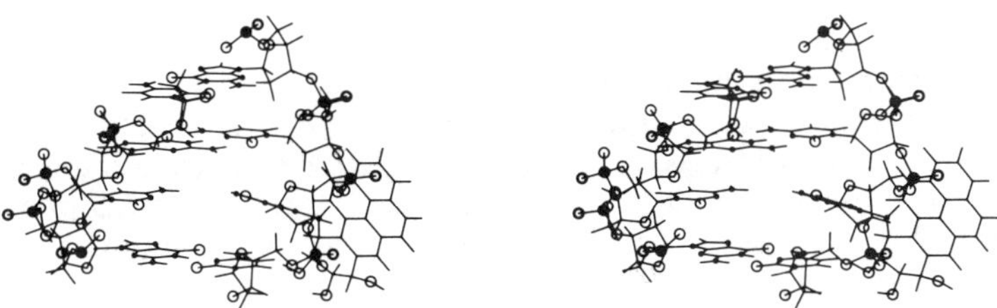

FIGURE 5. Stereographic projection of the externally bound BPDE-DNA adduct. The pyrene moiety lies in the major groove. The conformations of the various residues are those of B-DNA with the exception of the host guanine which resides in the *syn* conformation with $\chi_{syn} = \chi_{\text{B-DNA}} + 200°$.

not all molecules which intercalate exhibit activity, the connection between these antitumor agents, the intercalation process and the ability to inhibit nucleic acid synthesis suggests that a detailed molecular study of the binding process can be made. To date no procedure has been proposed that leads to the design of an antitumor agent. However, molecules that intercalate can be designed to fit into receptor sites in DNA in anticipation that some may exhibit antitumor activity.

The modeling procedure can predict binding orientations. The fit of 9-aminoacridine,[41] proflavin,[41,42] ethidium,[41,43–45] and daunomycin[46,47] in several

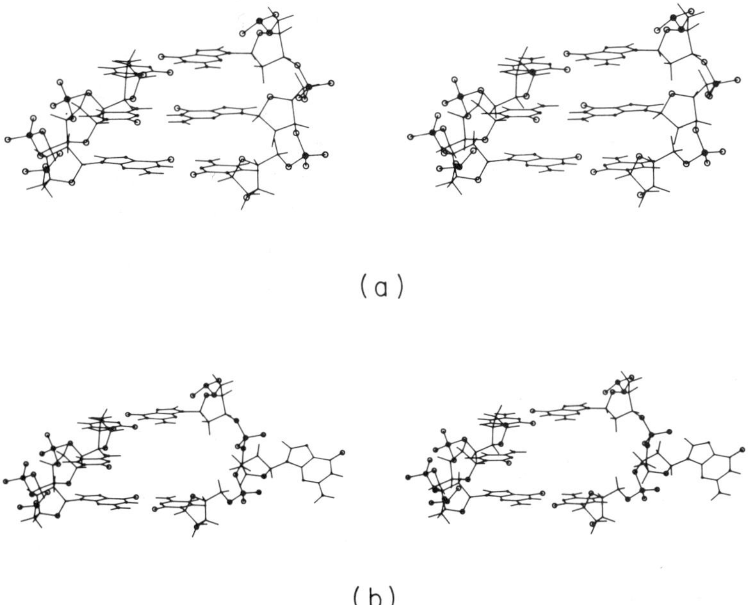

(a)

(b)

FIGURE 6. Normal **(a)** and crankshaft-altered **(b)** B-DNA. The crankshaft of **(b)** is from 03′ to P of the central nucleotide of the right-hand strand. The crankshaft rotation is 180° in this example.

model intercalation sites has been performed in several laboratories. The crystal structure of the dinucleoside monophosphate dimer duplex[48] has been used to model complex formation[44,45] as well as theoretically determined dimer duplex units.[47] End effects have been included by increasing the length of the polymer duplex about the binding site. A dinucleoside triphosphate minihelix[30,43] and a tetramer duplex[50] restrict the conformational space of the site. By assuming that the intercalation site can be inserted into a double helix, intercalation sites in a tetramer duplex have been constructed with the constraint that the end base pairs have sugar puckers and orientations of B-DNA.[42,51,52] With all of these model intercalation sites, the general features of binding have been obtained from theoretical calculations. We have shown that the conformational limitations imposed by the constraint that the site fit into a double helix, such as B-DNA, yield three families of intercalation sites. These receptor sites occur when the DNA unwinds by 4°–11°, 18°–22° and 24°–29° to yield intercalation sites I, II and III,[51,52] respectively, and this agrees with the extent of unwinding observed with closed circular DNA.[3,52]

To determine the optimum binding orientations of a molecule, the following calculations are performed. First, the minimum energy conformations of the free molecule are obtained. The molecule with conformational energies within 5–10 kcal of the global minimum is inserted into intercalation sites I, II and III. By starting with different initial orientations, the molecule is adjusted to fit into each of these receptor sites. The global minimum is chosen as the optimum fit of the molecule in the intercalation complex, and the site with lowest energy is assumed to represent the extent that the molecule unwinds the DNA. This procedure was used to predict the structure of lucanthone intercalated into DNA.[53] The experimental NMR measurements of lucanthone with poly(dA-dT)•poly(dA-dT) are in good agreement.[54]

To illustrate the intercalation process, the optimum binding orientations of molecules in DNA, and an approach to the design of intercalating agents, a systematic study is represented for a series of substituted anthraquinones. The substituent chosen is a protonated diethylaminoethylamino group.

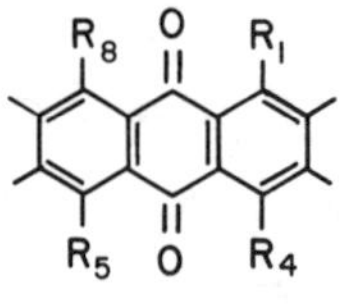

$$R = \overset{+}{N}H(CH_2CH_2)NH(CH_2CH_3)_2$$

The anthraquinones have been selected for study because several of them are not only active antitumor agents, but also intercalating agents.[55–59] Unfortunately, a large number do not exhibit activity.[60] To study the relationship between structure and activity, the model compounds are selected. Binding of these large substituents in the minor (m) and/or major (M) grooves is anticipated to yield insight into the overall orientation of the anthraquinone moiety in the receptor site, and the selection of a particular site.

Results for binding of 1-mono and 1,4- and 1,8-bis substituted anthraquinones (AQ) will be used to illustrate how the substituents control the orientation of the chromophore, how symmetry may be used to add substituents in strategic positions to enhance binding, and how substituents may select certain bases. In these studies, the substituents are found to be directed along the DNA backbone both above and below the plane. This is included in the notation. For example $18AQ(+Z, -Z)$ indicates that the 1 and 8 substituents are directed above $(+Z)$ and below $(-Z)$ the AQ plane, respectively.

The simplest system to study theoretically is anthraquinone itself. The global minimum occurs for maximum overlap with the adjacent base pairs. The short axis lies on the pseudodyad axis of the duplex. If one protonated amino group is added to yield $1AQ(+Z)$, then the global minimum occurs without disrupting this anthraquinone moiety, as illustrated in FIG. 7. This complex is particularly interesting because of its symmetry. It is analogous to lucanthone, which has been discussed extensively.[53] Rotation about this short axis places the amino substituent in the minor groove to proceed down the DNA. In each of the four symmetric DNA units such as ↑T•A,A•T↓, this rotation transforms the $1AQ(+Z)$ into an identical environment, both geometrically and energetically. In each of the ten remaining asymmetric DNA units such as ↑T•A,T•A↓, this rotation transforms the $1AQ(+Z)$ into a geometrically similar environment: it is identical only with respect to the phosphate backbones. Several observations can be made. First, the neutral anthraquinone moiety is sandwiched between adjacent base pairs to achieve maximum overlap. The van der Waals attractive energy is a maximum for

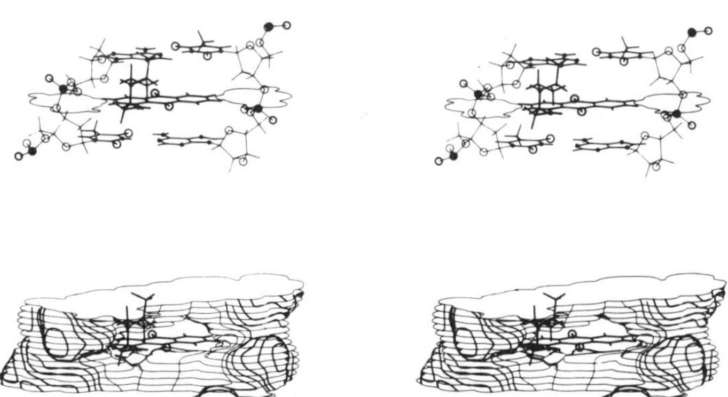

FIGURE 7. 1-diethylaminoethylamino anthraquinone $(+1)$ in intercalation site I of the ↑T•A,1AQ$(+Z)$,A•T↓ complex. The view is into the minor groove. A bond model with one steric contour and a steric model with electrostatic contours of negative (——), intermediate (——) and positive (━━) values.

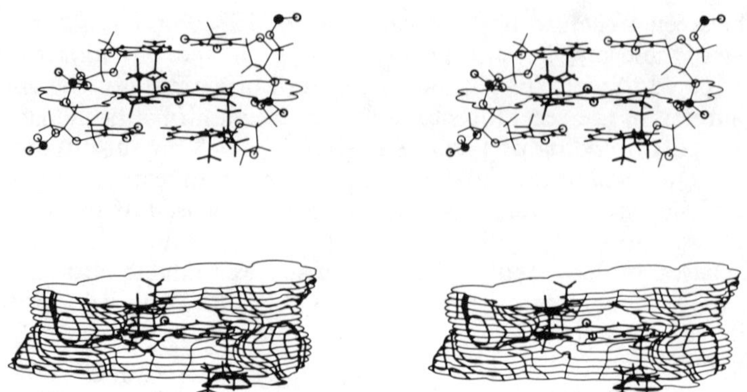

FIGURE 8. 1,8-bis(diethylaminoethylamino)anthraquinone ($+2$) in intercalation site I of the ↑T•A,18AQ($+Z,-Z$),A•T↓ complex. The view is into the minor groove.

this orientation. Second, the protonated terminal amino nitrogen resides in the most negative electrostatic region of the DNA as observed in the steric and electrostatic graphics presentation. Third, the global minimum energy conformation of the free 1AQ($+Z$) is approximately the same as that found in the intercalation complex. Therefore, the van der Waals and Coulomb interactions enhance the binding process, whereas the change in conformation energy during intercalation, ΔC, is nearly zero. Fourth, the optimum binding orientation of the anthraquinone AQ remains virtually unaltered upon substitution by this amino substituent. Therefore, the total binding energy is approximately a sum of binding for the AQ and the substituents.

One can take advantage of this symmetry by adding an identical amino group to produce the 1,8-bis(diethylaminoethylamino)-anthraquinone, 18AQ($+Z,-Z$). The optimum binding orientation of the anthraquinone

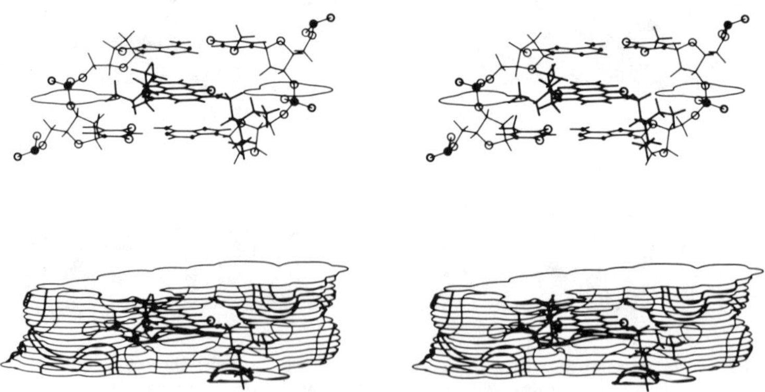

FIGURE 9. 1,4-bis(diethylaminoethylamino)anthraquinone ($+2$) in intercalation site III of the ↑T•A,14AQ($-Z,+Z$),A•T↓ complex. The view is into the minor groove.

moiety is the same as obtained for 1AQ(+ Z). It is shown in Fig. 8. The enhancement of binding results by placing a second protonated amino group on the DNA in exactly the same manner as obtained by a twofold rotation of 1AQ(+ Z). The dyad symmetry is apparent. The two-terminal protonated amino groups proceed along the antiparallel strands of the DNA and lie in the respective negative electrostatic regions. Once again, the total energy is approximately a sum of that found for the AQ moiety and the two amino substituents.

Another derivative occurs in which the 1- and 4-positions are substituted to yield 14AQ. There are two possibilities for binding orientations of the AQ moiety. The anthraquinone moiety is inserted through the intercalation site and the amino substituents are allowed to bind along the phosphate backbone analogous to 18AQ(+ Z, − Z). This 14AQ(− Z, + Z) example, shown in Fig. 9, yields a complex with an energy approximately 30 kcal higher than the 18AQ(+ Z, − Z). About half of this energy difference results from the reduced stacking energy of the anthraquinone moiety, and the remaining half is distributed between reduced Coulomb and steric interactions.

To maximize the stacking energy of the anthraquinone moiety, the substituents must be placed in the minor (m) and major (M) grooves respectively. This is shown in Fig. 10. Maximum overlap occurs between the anthraquinone moiety and adjacent base pairs with the amino groups bound in the

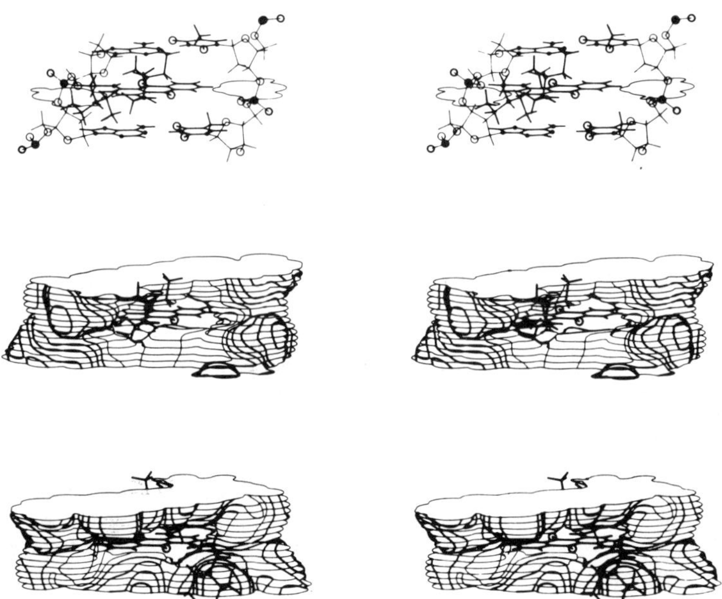

FIGURE 10. 1,4-bis(diethylaminoethylamino)anthraquinone (+ 2) in intercalation site I of the ↑A•T,14AQ(+ Z, − Z),A•T↓ complex. The views are into the minor groove (*upper and central*) and major groove (*lower*). The substituents straddle the grooves of the DNA.

TABLE 1. Relative Minimum Binding Energies of Substituted Anthraquinones Shown in FIGS. 7, 8 and 10[a]

	ΔE^I_{DNA}	$\delta\varepsilon^I_{1AQ(+Z)}$	$\delta\varepsilon^I_{18AQ(+Z,-Z)}$	$\delta\varepsilon^I_{14AQ(+Z,-Z)}$
$\uparrow$A•T$\downarrow$ / $\uparrow$T•A$\downarrow$	19.1	0.0	0.0	8.8
$\uparrow$A•T$\downarrow$ / $\uparrow$C•G$\downarrow$	20.0	0.5	4.0	6.5
$\uparrow$C•G$\downarrow$ / $\uparrow$T•A$\downarrow$	23.8	1.8	4.0	7.5
$\uparrow$C•G$\downarrow$ / $\uparrow$C•G$\downarrow$	24.7	2.4	7.9	7.1
$\uparrow$G•C$\downarrow$ / $\uparrow$T•A$\downarrow$	20.0	3.1	4.0	12.6
$\uparrow$G•C$\downarrow$ / $\uparrow$C•G$\downarrow$	20.5	3.3	7.7	13.2
$\uparrow$A•T$\downarrow$ / $\uparrow$G•C$\downarrow$	23.8	5.8	4.0	2.3
$\uparrow$A•T$\downarrow$ / $\uparrow$A•T$\downarrow$	25.3	6.9	8.7	0.0
$\uparrow$T•A$\downarrow$ / $\uparrow$T•A$\downarrow$	25.3	8.0	8.7	11.7
$\uparrow$C•G$\downarrow$ / $\uparrow$A•T$\downarrow$	29.6	8.3	11.5	3.6
$\uparrow$T•A$\downarrow$ / $\uparrow$C•G$\downarrow$	26.2	8.7	13.2	16.2
$\uparrow$G•C$\downarrow$ / $\uparrow$G•C$\downarrow$	24.7	8.8	7.9	5.4
$\uparrow$C•G$\downarrow$ / $\uparrow$G•C$\downarrow$	30.7	9.8	9.2	2.9
$\uparrow$G•C$\downarrow$ / $\uparrow$A•T$\downarrow$	26.2	9.9	13.2	6.1
$\uparrow$T•A$\downarrow$ / $\uparrow$A•T$\downarrow$	29.1	12.4	14.5	8.3
$\uparrow$T•A$\downarrow$ / $\uparrow$G•C$\downarrow$	29.6	13.3	11.5	7.8
$(\varepsilon^I_{MOLC})min$	—	−197.5	−375.7	−347.3

[a] The opening energies, ΔE^I_{DNA}, and binding energies, ΔE^I_{MOLC}, are combined to yield the complex energies $\Delta\varepsilon^I_{MOLC}$ defined in Eq. **(1)**. The relative complex energy is $\delta\varepsilon^I_{MOLC} = \varepsilon^I_{MOLC} - (\varepsilon^I_{MOLC})\ min$.

most negative electrostatic region in the minor and major grooves respectively. The binding energy of this 14AQ($+Z, -Z$)-DNA intercalation complex is approximately the same as the 18AQ($+Z, -Z$)-DNA complex.

Base sequence specificity toward T•A has been observed experimentally in daunomycin[61–63] and obtained by theoretical calculations in our laboratory.[46] The anthraquinone derivatives are particularly interesting because the addition of substituents enhances selectivity for T•A. In TABLE 1, the complex energies ΔE^v_{MOLC} of 1AQ($+$Z), 18AQ($+$Z, $-$Z) and 14AQ($+$Z, $-$Z) are reported for the optimum intercalation complexes. There are two important contributions in the binding process. Both assist in the selection process for the 1AQ($+$Z) and 18AQ($+$Z, $-$Z) but not for 14AQ($+$Z, $-$Z). The energy required to open the DNA to an intercalation site v, ΔE^v_{DNA}, favors (pyrimidine)p(purine) sequences. The protonated substituents bind to the negatively charged phosphate backbones, and, in particular, the 2′-deoxyribo 5′-monophosphate portion of the nucleotide containing the base adenine is most negative.[46] The selectivity of the ↑T•A,A•T↓ sequence increases with 18AQ($+$Z, $-$Z) over 1AQ($+$Z) as seen by examining $\delta\varepsilon$ in TABLE 1. Symmetry in binding is responsible for the enhancement of the binding energy.

The situation is quite different for 14AQ($+$Z, $-$Z). Reorientation of the anthraquinone moiety but, especially, the change in the direction of binding of the amino substituents along the backbone yields a different order. In this case the opening energies, $\Delta E^{\text{I}}_{\text{DNA}}$, and the binding energies, $\Delta E^{\text{I}}_{\text{14AQ}}$, do not simultaneously enhance the complex energies, nor do they favor the (pyrimidine)p(purine) sequences. The most favored is ↑A•T,A•T↓ by 2.3 kcal. Both amino substituents bind along the strand containing adenine of dApdA in the optimum complex. This is illustrated in FIG. 10.

The predictions which we make for base specificity depend on the strand to which the substituents bind. For the 18AQ, dTpdA base specificity is obtained while 14AQ selects dApdA. Both, however, favor the T•A base pairs. In addition, these as well as the 1AQ exhibit optimum binding in intercalation site I, and we predict an unwinding angle of 4°–11°.

COMPUTER GRAPHICS

To illustrate the conformations of the complexes and the binding process, a bond and a space-filling representation has been used.[24,25] The steric contours are generated in planes of constant z with 6–14 potential $U(x,y)$. Rather than generating a grid of potential points and interpolating, the derivatives dy/dx and (d^2y/dx^2) are evaluated on the contour $U = 0$ and the Taylor's expansion

$$\Delta y = (dy/dx)\Delta x + (d^2y/dx^2)\Delta x^2/2 \tag{3}$$

is used to step around this constant contour. When given electrostatic values are encountered on this steric contour, they are saved and connected after the entire steric surface has been drawn.[26] To aid in the presentation the hidden lines are removed. This feature has the advantage that portions of a molecule that penetrate the space occupied by the receptor site are removed. Poor contacts are represented by the penetration of the molecule into the DNA by a removal of those portions of the molecule that penetrate the steric surface. Poor fits can be used to assist in the adjustment of the molecule to the receptor site or to expose a poor model. The sequence which results when the contours are drawn is illustrated in FIG. 11 for 18AQ($+$Z, $-$Z).

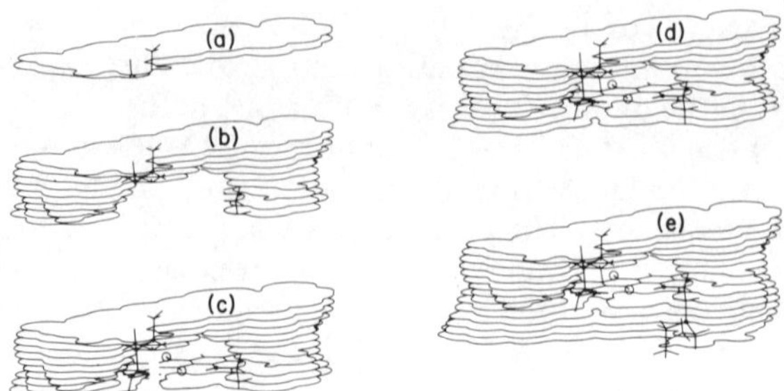

FIGURE 11. Stepwise procedure illustrating the construction of $18AQ(+Z, -Z)$ as each contour is drawn. **(a)** Portions of bonds placed above the first and between the first two contours, **(b)** continuation of drawing of contours, **(c)** drawing the molecule as the intercalation hole evolves, **(d)** completion of the molecule in the intercalation hole, and **(e)** completion of the molecule with the remaining contours.

The process of creating a model for the interaction of a molecule with a receptor site is performed as follows. A systematic analysis yields many geometrical structures with the requirement that base pairs and the bases themselves have a given orientation. This is the time-consuming step. Then the acceptable ones are obtained by requiring that each structure has favorable nonbonded contacts between atoms. Thus, both geometric and energetic requirements are considered. Families of conformations are classified from which the optimum ones are selected as trial receptor sites. The interactive mode is used to adjust a molecule to fit the receptor site and to perform adjustments of the DNA such as crankshaft motion and rotations about the glycosidic bonds. A bond model is used for this process. Once the contours are obtained, the steric-electrostatic contour presentation requires about 15 s per projection if hidden lines are removed and about 1 s if all lines are drawn. At present the contours are a useful diagnostic tool and convenient to present the spatial and electrostatic characteristics of receptor sites.

The combined developments in the generation of nucleic acid structures, the binding of molecules to receptor sites and computer graphics techniques provide the necessary tools to assist the chemist in the computer-assisted analysis of the interaction of antitumor agents and carcinogens with DNA.

REFERENCES

1. LERMAN, L. 1961. J. Mol. Biol. **3**: 18–30.
2. OSTER, G. 1951. Trans. Far. Soc. **47**: 660–666.
3. GALE, E. F., E. CUNDLIFFE, P. E. REYNOLDS, M. H. RICHMOND & M. J. WARING. 1981. Inhibitors of nucleic acid synthesis. *In* The Molecular Basis of Antibiotic Action. 2d ed. Wiley. New York. 258–401.
4. GEACINTOV, N. E., H. YOSHIDA, V. IBANEZ & R. G. HARVEY. 1982. Biochemistry **21**: 1864–1869.
5. WEINSTEIN, I. B., A. M. JEFFREY, K. W. JENNETTE, S. H. BLOBSTEIN, R. G. HARVEY, C. HARRIS, H. AUTRUP, H. KASAI & K. NAKANISHI. 1976. Science **193**: 592–595.

6. MAGER, R., E. HUBERMAN, W. S. K. YANG, H. V. GELBOIN & L. SACHS. 1977. Int. J. Cancer **19**: 814–817.

7. JEFFREY, A. M., I. B. WEINSTEIN, K. W. JENNETTE, K. GRZESKOWIAK, K. NAKANISHI, R. G. HARVEY, H. AUTRUP & C. HARRIS. 1977. Nature (London) **269**: 348–350.

8. MOORE, P. D., M. KOREEDA, P. G. WISLOCKI, A. H. LEV, H. YAGI & D. M. JERINA. 1978. Am. Chem. Soc. Symp. Ser. **44**: 127–154.

9. KOREEDA, M., P. D. MOORE, H. YAGI, H. J. C. YEH & D. M. JERINA. 1976. J. Am. Chem. Soc. **98**: 6720–6722.

10. JENNETTE, K. W., A. M. JEFFREY, S. H. BLOBSTEIN, F. A. BELAND, R. G. HARVEY & I. B. WEINSTEIN. 1977. Biochemistry **16**: 932–938.

11. MEEHAN, T., K. STRAUB & M. CALVIN. 1977. Nature (London) **269**: 725–727.

12. JEFFREY, A. M., K. GRZESKOWIAK, I. B. WEINSTEIN, K. NAKANISHI, P. ROLLER & R. G. HARVEY. 1979. Science **206**: 1309–1311.

13. IVANOVIC, V., N. E. GEACINTOV, H. YAMASAKI & I. B. WEINSTEIN 1978. Biochemistry **17**: 1597–1603.

14. KOREEDA, M., P. D. MOORE, P. G. WILSON, W. LEVIN, A. H. CONNEY, H. YAGI & A. M. JEFFREY. 1978. Science **199**: 778–781.

15. JEFFREY, A. M., K. W. JENNETTE, S. H. BLOBSTEIN, I. B. WEINSTEIN, F. A. BELAND, R. G. HARVEY, H. KASAI, I. MIRURA & K. NAKANISHI. 1976. J. Am. Chem. Soc. **98**: 5714–5715.

16. MOORE, P. D., M. KOREEDA, P. G. WISLOKI, W. LEVIN, A. H. CONNEY, H. YAGI & D. M. JERINA. 1977. Drug Metabolism Concepts. D. M. Jerina, Ed.: 127–154. Am. Chem. Soc. Washington, D. C.

17. GAMPER, H. B., S. S. C. TUNG, K. STRAUB, J. C. BARTHOLOMEW & M. CALVIN. 1977. Science **197**: 671–674.

18. BORGAN, A., H. DARVEY, N. CATAGNOLI, T. T. CROCKER, R. E. RASMUSSEN & I. Y. WANG. 1973. J. Med. Chem. **16**: 502–506.

19. SIMS, P., P. L. GROVER, A. SWAISLAND, K. PAL & A. HEWER. 1974. Nature (London) **252**: 326–327.

20. DAUDEL, P., M. DUQUESNE, P. VIGNY, P. L. GROVER & P. SIMS. 1975. FEBS Lett. **57**: 250–253.

21. IVANOVIC, V., N. E. GEACINTOV & I. B. WEINSTEIN. 1976. Biochem. Biophys. Res. Commun. **70**: 1172–1179.

22. MILLER, K. J. 1979. Biopolymers **18**: 959–980.

23. TAYLOR, E. R., K. J. MILLER & A. J. BLEYER. 1983. J. Biomolecular Structure and Dynamics **1**: 883–904.

24. MILLER, K. J., P. J. KOWALCZYK & W. SEGMULLER. 1981. *In* Polynuclear Aromatic Hydrocarbons: Sixth International Symposium on Physical and Biological Chemistry. M. Cooke & A. Dennis, Eds.: 529–535. Battelle Press. Columbus, Ohio.

25. MILLER, K. J., P. KOWALCZYK, W. SEGMULLER & G. WALKER. 1983. J. Comp. Chem. **4**: 366–378.

26. MILLER, K. J. & P. J. KOWALCZYK. 1984. J. Comp. Chem. **5**: 89–103.

27. OLSON, W. K. & P. J. FLORY. 1972. Biopolymers **11**: 25–56.

28. ZHURKIN, V. B., Y. P. LYSOV & V. I. IVANOV. 1978. Biopolymers **17**: 377–412.

29. SMITH, P. J. C. & S. ARNOTT. 1978. Acta Crystallogr. **A34**: 3–11.

30. ORNSTEIN, R. & R. REIN. 1979. Biopolymers **18**: 2821–2847.

31. JERINA, D. M. & J. W. DALY. 1974. Science **185**: 573–581.

32. BROOKES, P. & P. D. LAWLEY. 1978. Nature (London) **202**: 781–784.

33. MILLER, J. A. 1970. Cancer Res. **30**: 559–576.

34. GELBOIN, H. V., N. KINOSHITA & F. J. WIEBEL. 1972. Fed. Proc. Fed. Am. Soc. Exp. Biol. **31**: 1298–1309.

35. GEACINTOV, N. E., A. GAGLIANO, V. IVANOVIC & I. B. WEINSTEIN. 1978. Biochemistry 17: 5256–5262.
36. HOGAN, M. E., N. DATTAGUPTA & J. P. WHITLOCK, JR. 1981. J. Biol. Chem. 256: 4505–4513.
37. GAMPER, H. B., K. STRAUB, M. CALVIN & J. C. BARTHOLOMEW. 1980. Proc. Nat. Acad. Sci. U.S.A. 77: 2000–2004.
38. BELAND, F. A. 1978. Chem.-Biol. Interact. 22: 329–339.
39. JEFFREY, A. M., T. KINOSHITA, R. M. SANTELLA, D. GRUNBERGER, L. KATZ & I. B. WEINSTEIN. 1980. Jerusalem Symposium on Quant. Chem. and Biochem., vol. 13. B. Pullman, P. O. P. Tso & H. Gelboin, Eds.: 565–579. Jerusalem, Israel.
40. MILLER, K. J., J. J. BURBAUM & J. DOMMEN. 1981. *In* Polynuclear Aromatic Hydrocarbons: Sixth International Symposium on Physical and Biological Chemistry. M. Cooke & A. Dennis, Eds.: 515–528. Battelle Press. Columbus, Ohio.
41. MILLER, K. J., R. BRODZINSKY & S. HALL. 1980. Biopolymers 19: 2091–2122.
42. ALDEN, C. J. & S. ARNOTT. 1977. Nucl. Acids Res. 4: 3855–3861.
43. ORNSTEIN, R. L. & R. REIN. 1979. Biopolymers 18: 1277–1291.
44. PACK, G. R. & G. H. LOEW. 1978. Biochim. Biophys. Acta 519: 163–172.
45. NUSS, M. E., F. J. MARSH & P. A. KOLLMAN. 1979. J. Amer. Chem. Soc. 101: 825–833.
46. NEWLIN, D. D., K. J. MILLER & D. F. PILCH. 1984. Biopolymers 23: 139–158.
47. NAKATA, Y. & A. J. HOPFINGER. 1980. Biochem. Biophys. Res. Commun. 95: 583–588.
48. TSAI, C. C., S. C. JAIN & H. M. SOBELL. 1977. J. Mol. Biol. 114: 301–315.
49. BERMAN, H. M., S. NEIDLE & R. K. STODOLA. 1978. Proc. Natl. Acad. Sci. U.S.A. 75: 828–832.
50. BERMAN, H. M. & S. NEIDLE. 1979. *In* Stereodynamics of Molecular Systems. R. H. Sarma, Ed.: 367–382. Pergamon. New York.
51. MILLER, K. J. & J. F. PYCIOR. 1979. Biopolymers 18: 2683–2719.
52. MILLER, K. J. 1981. *In* Proceedings of the Second SUNYA Conversation in the Discipline Biomolecular Stereodynamics, vol. II. R. H. Sarma, Ed.: 469–486. Adenine Press. New York.
53. MILLER, K. J., M. LAUER & S. ARCHER. 1980. Inter. J. Quant. Chem: Quantum Biol. Sym. 7: 11–34.
54. SARMA, M. H., C. K. MITRA, R. H. SARMA, K. J. MILLER & S. ARCHER. 1980. Biochem. Biophys. Res. Commun. 94: 1285–1295.
55. JOHNSON, R. K., R. K.-Y. ZEE-CHENG, W. W. LEE, E. M. ACTON, D. W. HENRY & C. C. CHENG. 1979. Cancer Treatment Reports 63: 425–439.
56. ZEE-CHENG, R. K.-Y. & C. C. CHENG. 1978. J. Med. Chem. 21: 291–294.
57. ZEE-CHENG, R. K.-Y., E. G. PODREBARAC, C. S. MENOM & C. C. CHENG. 1979. J. Med. Chem. 22: 501–505.
58. CHENG, C. C., G. ZBINDEN & R. K.-Y. ZEE-CHENG. 1979. J. Pharm. Sci. 68: 393–396.
59. MURDOCK, K. C., R. G. CHILD, P. F. FABIO, R. B. ANGIER, R. E. WALLACE, F. E. DURR & R. V. CITARELLA. 1979. J. Med. Chem. 22: 1024–1030.
60. DRISCOLL, J. S., G. F. HAZARD, JR., H. B. WOOD & A. GOLDIN. 1974. Cancer Chemother. Rep. 4: 1–362.
61. CHANDRA, P., F. UNINO, A. GOTZ, D. GERIEKI, R. THORBECK & A. DIMARCO. 1972. FEBS Lett. 21: 264–268.
62. GRAVES, D. E. & T. R. KRUGH. 1983. Biochemistry 22: 3941–3947.
63. CHAIRES, J. B. 1983. Biochemistry 22: 4204–4211.

Molecular Dynamics and Minimum Energy Conformations of GnRH and Analogs

A Methodology for Computer-aided Drug Design[a]

R. S. STRUTHERS,[b,c] J. RIVIER,[c] AND A. T. HAGLER[b,c]

[b]*The Agouron Institute*
La Jolla, California 92037

[c]*Peptide Biology Laboratory*
The Salk Institute
La Jolla, California 92138

Gonadotropin-releasing hormone (GnRH) is one of the hypothalamic releasing factors that control pituitary hormone secretion. Under control of higher central nervous system centers, GnRH is released by neuronal cells of the hypothalamus into the primary plexus of the hypothalamus-pituitary portal system. It then travels with the portal blood down the infundibular stalk to the secondary plexus in the anterior pituitary where it causes the release of the gonadotropins, follicle-stimulating hormone (FSH) and luteinizing hormone (LH).

The gonadotropins, in turn, have a wide range of effects controlling reproductive functions in both male and female, including steroid production and metabolism, gonad growth and development, spermatogenesis, oogenesis, and ovulation. More recently, GnRH has been found to have a number of extrapituitary actions including direct effects on gonadal functions.[1,2] Long-term administration of pharmacologic doses of GnRH and GnRH agonist have been shown to inhibit a number of male or female reproductive functions.[1,2]

Sequence Activity Studies

Since the isolation and sequence determination of porcine[3] and ovine[4] GnRH (see below), over two thousand analogs of the peptide have been synthesized and biologically evaluated. Among these, competitive receptor antagonists to GnRH were developed that have been shown to inhibit both ovulation and spermatogenesis and have therefore elicited an intense interest as possible contraceptive agents.[5] Some key analogs in the development of GnRH antagonists are given in TABLE 1.

[a] This work was supported in part by contract no. NOI–HD–2–2807 and grant no. HD–13527 from the National Institutes of Health, and grant no. PCM–8204908 from the National Science Foundation.

Traditionally, the design of peptide analogs has involved sequence activity studies in order to determine which residues are necessary for binding and transduction of the biological message,[6] and also to optimize steric and electronic parameters of these residues.[7] The initial discovery leading to the development of peptide antagonists to GnRH was that des-His²-GnRH was found to exhibit minimal agonist activity and could inhibit the activity of GnRH *in vitro*.[8] It was later shown that this class of analogs competes against GnRH for high-affinity binding sites on pituitary cells.[9]

Following this lead, D-Phe was substituted for L-His at position 2 and shown to possess enhanced antagonist properties.[10] Substitution of D-amino acids at position 6, especially those with aromatic sidechains, resulted in agonists with enhanced activity[11] and this effect proved to be transferable to the antagonist series.[6,12] The systematic optimization of residues at positions 1 and 2 in the basic structure containing D-Trp at positions 3 and 6 led to the very potent antagonists [Ac-Δ³,⁴-Pro¹, D-Phe²(4-Cl), D-Trp³,⁶]-GnRH which prevents ovulation in eight out of eight female rats when 7.5 μg of material is administered at noon on the day of proestrus.[7]

More recently, it was discovered that a D-Arg at position 6, in combination with a Ac-D-Phe(4Cl) at position 1, results in somewhat more potent analogs,[13] whereas introduction of an even more hydrophobic residue, such as D-2-naphthylalanine[D-Nal(2)], at position 1 further increased potency. Finally, extended duration of action could be obtained by substitution on the guanidino function of D-Arg⁶ with ethyl groups.[14]

The Need for Conformation Activity Studies

In order to understand the nature of the structural requirements for the high binding affinity of GnRH and its analogs, it is not sufficient to consider only the number and nature of the functional groups, as has been done in sequence activity studies. It is incumbent upon us to consider also the three-dimensional positioning of these groups. Presumably it is the relative orientation of the sidechains that is responsible for specific binding. Very subtle changes in structure can often produce profound changes in biological activity. For example, replacing Gly⁶ with L-alanine results in an analog with low potency, whereas replacement with D-Ala results in an analog with a four-fold increase in potency.[15] This phenomenon of high sensitivity of biological

TABLE 1. Key Analogs in the Development of Potent GnRH Antagonists

| | | Potency | | |
	Sequence	$ICR_{50}{}^{a}$	A.O.A.[b]	Ref.
1.	GnRH pGlu-His-Trp-Ser-Tyr-Gly-Leu-Arg-Pro-Gly-NH$_2$			
2.	pGlu- -Trp-Ser-Tyr-Gly-Leu-Arg-Pro-Gly-NH$_2$	1000–4000	—	8
3.	pGlu-D-Phe-Trp-Ser-Tyr-Gly-Leu-Arg-Pro-Gly-NH$_2$	200^c	—	10
4.	D-pGlu-D-Phe-D-Trp-Ser-Tyr-D-Trp-Leu-Arg-Pro-Gly-NH$_2$	3.0	250 µg (0/10)	12
5.	Ac-$\Delta^{3,4}$-Pro-D-Phe(4Cl)-D-Trp-Ser-Tyr-D-Trp-Leu-Arg-Pro-Gly-NH$_2$	0.04–0.02	7.5 µg (0/10)	7
6.	Ac-D-Nal(2)-D-Phe(4F)-D-Trp-Ser-Tyr-D-Arg-Leu-Arg-Pro-Gly-NH$_2$	—	1.0 µg (0/10)	
7.	cyclo[$\Delta^{3,4}$-Pro-D-Phe(4Cl)-D-Trp-Ser-Tyr-D-Trp-NMe-Leu-Arg-Pro-β-Ala]	3.5	—	7

^{a} Potencies as ICR_{50} are expressed as the concentration ratios (antagonist/GnRH) required to reduce the amount of GnRH-induced LH secretion by 50%.

^{b} The peptides were injected subcutaneously into cycling rats at noon on the day of proestrus. Results for the given dosage are expressed in terms of the ratio of GnRH antagonist treated rats ovulating over vehicle-administered controls.

^{c} Minimum concentration ratio antagonizing GnRH response.

activity to minor alteration in structure implies highly specific recognition on the part of the receptor of a three-dimensional constellation of functionalities presented by the peptide.

In order to try to elucidate the conformational properties that may be underlying binding to the GnRH receptor, we have undertaken an extensive theoretical study of the conformation, energetics and dynamics of GnRH and several analogs. These properties can then be used as the basis for the design of new analogs for synthesis and biological evaluation, in order to test conformational hypotheses. The iterative application of this process of rational analog design should ultimately establish the binding conformation(s) of GnRH and its analogs, and hopefully lead to the development of useful therapeutic agents.

METHODS FOR SIMULATION OF GnRH

The first step in the application of this process is to gain a better understanding of the conformations accessible to the native GnRH decapeptide. This has been done by applying the modern technique of molecular dynamics, which allows us to follow the motions of a molecule and observe both vibrational fluctuations around a given conformational state and major transitions from one conformational state to another.[16]

Molecular Dynamics

The first step in performing a molecular dynamics simulation is specifying the energy surface of the system. This is done using a valence force field, as is given in (1), which takes into account all degrees of freedom, including bond lengths, bond angles, and torsion angles.

$$\begin{aligned}
V = {}& \Sigma\{D_b[1 - e^{-\alpha(b-b_0)}]^2 - D_b\} + \tfrac{1}{2} \Sigma K_\theta (\theta - \theta_0)^2 \\
& + \tfrac{1}{2} \Sigma K_\phi (1 + s \cos n\phi) + \tfrac{1}{2} \Sigma K_\chi \chi^2 \\
& + \Sigma\Sigma F_{bb'}(b - b_0)(b' - b_0') \\
& + \Sigma\Sigma F_{\theta\theta'}(\theta - \theta_0)(\theta' - \theta_0') + \Sigma\Sigma F_{b\theta}(b - b_0)(\theta - \theta_0) \\
& + \Sigma F_{\phi\theta\theta'} \cos \phi(\theta - \theta_0)(\theta' - \theta_0') + \Sigma\Sigma F_{\chi\chi'}\chi\chi' \\
& + \Sigma\varepsilon[2(r^*/r)^{12} - 3(r^*/r)^6] + \Sigma q_i q_j/r
\end{aligned} \tag{1}$$

There are two types of quantities represented in the equation: constants characteristic of the energy required to deform a given internal coordinate ($K_\theta \ldots F_{bb}$, $F_{\theta\theta} \ldots$) or characteristic of the strength of a given interatomic nonbonded interaction (r^*, E, q); and the internal coordinates specifying the geometry of the molecule (bond lengths, b; bond angles, θ; dihedral angles, ϕ; out of plane angles, χ; and nonbonded interatomic distances, r). From the energy surface, we can obtain the forces on the individual atoms by calculating the negative of the first derivatives of the energy with respect to the

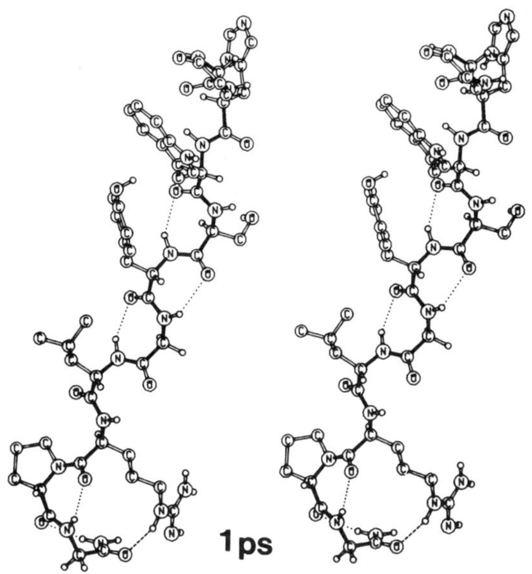

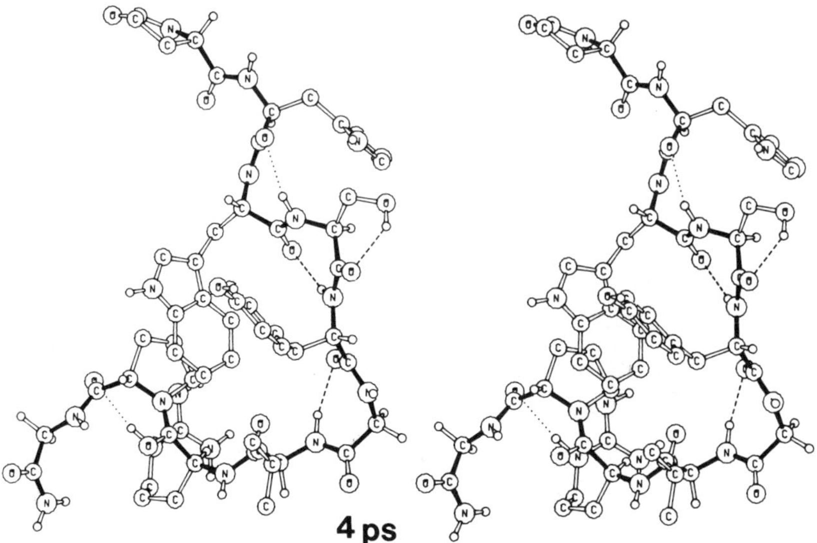

FIGURE 1. Selected minimum energy structures of GnRH derived by energy minimization of the instantaneous molecular dynamics configuration at the indicated elapsed time in the trajectory.

FIGURE 1 *(continued)*

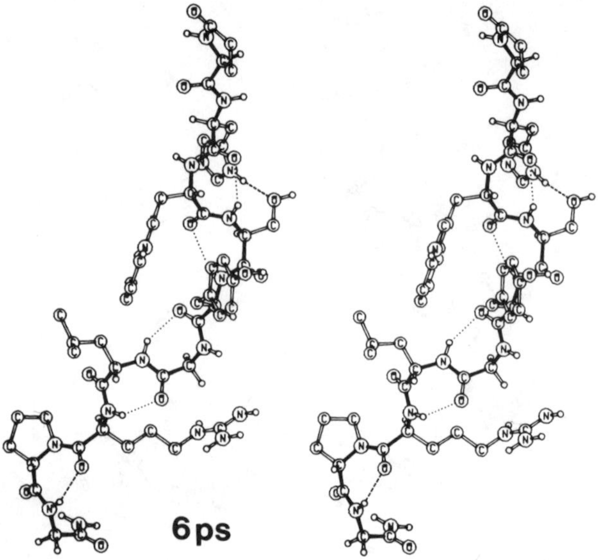

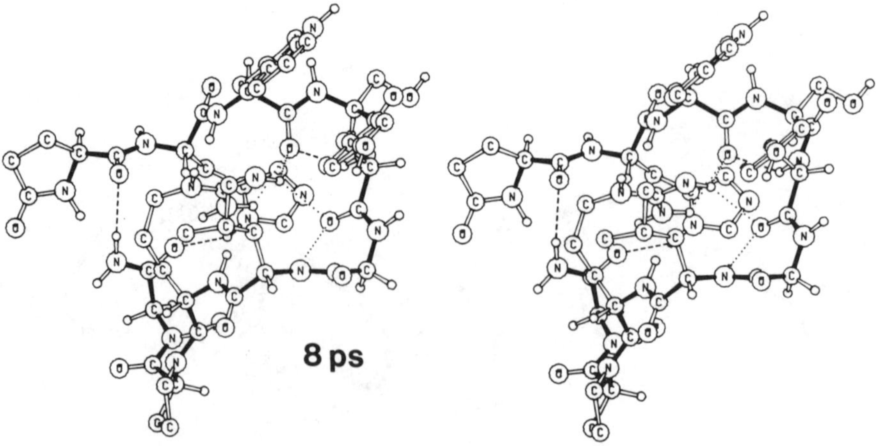

FIGURE 1 *(continued)*

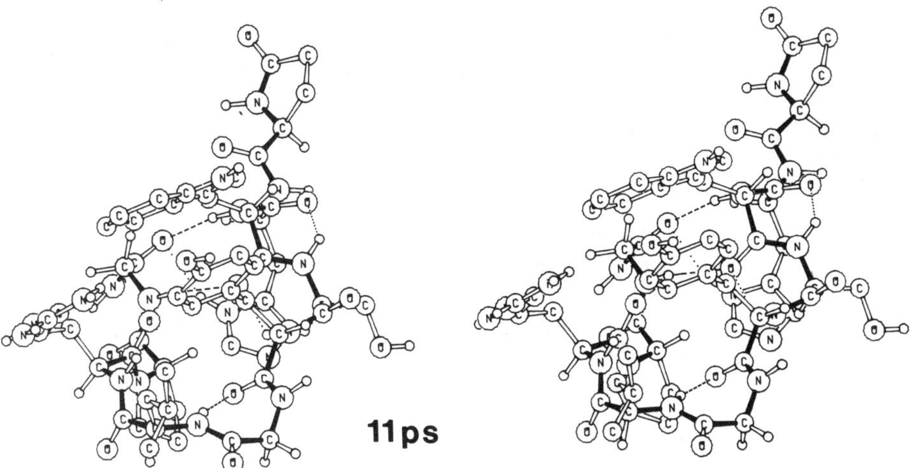

Cartesian coordinates of the atoms. Since we know the masses of the atoms, by substituting these forces into Newton's equation of motion ($F = ma$) we can integrate the accelerations forward in time to follow the trajectory of the atoms, giving us the molecular dynamics of the system.

Therefore, in order to calculate the molecular dynamics trajectory of GnRH, we require a set of potential parameters that allow for the calculation of the energy surface of the molecule, and a set of initial conditions for the system (*i.e.*, x, y and z coordinates for all atoms in the molecule). The potential parameters used in this study were taken primarily from those reported previously.[17-25] The initial coordinates for the GnRH molecular dynamics simulation were obtained by partial minimization of a fully extended structure (ϕ, ψ = -180, $+180$ for each residue except *p*Glu[1] and Pro[9]) in order to eliminate bad steric contacts.

Energy Minimization to Characterize Conformational States of GnRH

A 14 ps molecular dynamics simulation of the GnRH decapeptide was performed. At the beginning of the simulation, the molecule was in an extended conformation and over the course of this short simulation folded into more compact conformations, unfolded, then refolded into very compact structures. These different conformational states can be characterized by the local minimum energy conformation, about which the oscillations observed in molecular dynamics occur. The minimum energy conformations can be derived by minimizing the energy of the molecule with respect to all degrees of freedom in the molecule (*i.e.*, solving for the structure where the derivatives of the energy (1) with respect to the atomic coordinates are zero). The minimum energy structures about which fluctuations are taking place were found in this way using the instantaneous conformations at 1 ps intervals along the molecular dynamics trajectory as starting structures for minimization. A number of these minimum energy structures are shown in stereo ortep drawings in FIGURE 1. The total energy and energy components of all minima found are given in TABLE 2.

A DESCRIPTION OF MINIMUM ENERGY CONFORMATIONS OF GnRH

One Picosecond. As noted above, the starting structure for this molecular dynamic simulation was obtained by partial minimization of a fully extended conformation which resulted in small changes in backbone torsion angles of less than ~20°. After 1 ps of molecular dynamics simulation, the resulting minimum energy structure is still extended with most residues moving into the C_7^{eq} conformation (ϕ, ψ = $-80°$, $+80°$). This results in the formation of a number of hydrogen bonds about the residues Ser[4][Trp[3](CO) . . . (NH)Tyr[5]], Tyr[5][Ser[4](CO) . . . (NH)Gly[6]], Gly[6][Tyr[5](CO) . . . (NH)Leu[7]], Leu[7][Gly[6](CO) . . . (NH)Arg[8]], Pro[9][Arg[8](CO) . . . (NH)Gly[10]], and Gly[10] [Pro[9](CO) . . . (NH)amide] which are seen in FIGURE 1 as dotted or dashed lines. An interesting aspect of this structure is the amphipathic organization of the sidechains with pGlu[1], Trp[3], Tyr[5], Leu[7] and Pro[9] on the one side, and the hydrophilic sidechains of His[2], Ser[4], Arg[8], and the Gly[10] amide on the other. This reflects the alternating hydrophobic-hydrophilic sequence which is manifested structurally when the peptide exists in an extended structure.

TABLE 2. Total Energy and Energy Components of the Minimum Energy Structures Underlying the Dynamic Trajectory of GnRH

Time (ps)	Potential Energy		
	Valence[a]	Nonbond[a]	Total
0	120.9	-43.8	77.1 (20.8)
1	119.4	-42.3	77.1 (20.8)
2	125.2	-39.8	85.3 (29.0)
3	125.6	-55.2	70.4 (14.1)
4	123.3	-54.1	69.3 (13.0)
5	118.9	-46.7	72.1 (15.8)
6	118.6	-44.1	74.5 (18.2)
7	125.8	-51.2	74.6 (17.7)
8	130.7	-70.7	60.0 (3.7)
9	130.0	-69.7	60.3 (4.0)
10	128.5	-69.2	59.3 (3.0)
11	125.6	-66.5	59.1 (2.8)
12	125.3	-68.9	56.3 (0.0)

[a] The valence term represents deformations of internal coordinates, while the nonbond refers to Lennard-Jones and electrostatic interactions (see reference 16).

The potential energy of this conformation is still rather high, being ~20 kcal/mole above the lowest energy minimum found for GnRH in this study.

Although without its explicit inclusion into the calculation the effect of environment cannot be predicted, it is possible to speculate that conformations related to this one may be adopted when the peptide is exposed to amphipathic environments such as an air-water interface or the surface of a lipid membrane, in much the same way that corticotropin-releasing factor, which is generally disorganized in aqueous solution, readily assumes an amphipathic secondary structure in response to these amphipathic environments.[26]

Four Picoseconds. After 4 ps of simulation the molecule has adopted a folded conformation. The mainchain hydrogen bonds about Ser[4] and Gly[6] have been maintained with the addition of a C_7 hydrogen bond at Trp[3][His[2](CO) . . . (NH)Ser[4]], while those about Tyr[5] and Leu[7] have been broken, corresponding to the adoption of an extended conformation (ϕ, ψ = $-160°$, $+160°$) by these residues. These conformational transitions of Tyr[5] and Leu[7] result in a turn, or perhaps more properly a "kink," in the chain being formed at Gly[6], allowing "distant" residues to interact.

An interesting feature of this conformation is the "hydrophobic core" formed at the inside of the turn by the mutual interactions of the sidechains of Trp[3], Tyr[5], Leu[7], and Pro[9] which were brought into contact by this turn at Gly[6]. This is an appealing structural element to invoke in considering stable structure in an aqueous environment. In fact, this conformation may provide an additional explanation for the nuclear Overhauser observations made by Sprecher and Momany[27] on GnRH in D_2O.

Six Picoseconds. After 6 ps of simulation the molecule has returned to a more extended structure. Leu[7] has returned to a C_7^{eq} (ϕ, ψ = -84, 83) conformation restoring the hydrogen bond absent in the minimum at 4 ps. This change in the conformation of Leu[7] has the effect of moving the carboxy terminal residues away from the rest of the molecule, and disrupting the long-range interactions present in the previous structure. Previous conformations exhibited sidechain to mainchain hydrogen bonds, such as the hydroxyl of Ser 4 hydrogen bonding to its own carbonyl and the Arg[8] guanidinium hydrogen bonding to the carbonyl of Pro[9], and here we observe a sidechain to sidechain hydrogen bond formed between the His[2](NH^{im}) and the Ser[4] hydroxyl oxygen.

This structure is the highest energy link in a chain of intermediate conformations between the relatively low-energy conformation at 4 ps and the still lower energy structures beginning at 8 ps. It serves to demonstrate once again the power of dynamics techniques to sample conformational transitions to higher energy states. The potential energy of this conformation is some 5 kcal/mole higher than that after 4 ps. In the minimun at 4 ps the Arg[8] sidechain is hydrogen bonded to the Pro[9] carbonyl, and is in effect pointing toward the carboxy terminus. At 6 ps, this hydrogen bond has been broken and the Arg[8] sidechain has moved more toward the amino terminus. The movement of the Arg[8] is concomitant with a transition in Leu[7] from C_7^{eq} (ϕ, ψ = $-80°$, $+80°$) to α' (ϕ, ψ = $-130°$, $-60°$). This bridging action of the Arg[8] sidechain to residues preceding it in the sequence appears to be "nucleating" the transition to the very compact structures that follow.

Eight Picoseconds. A major change in shape has occurred to reach the very compact minimum energy structure after 8 ps. The guanidinium sidechain of Arg[8] is buried in the center of an unorthodox turn involving five residues (Trp-Ser-Tyr-Gly-Leu) and is involved in an extensive hydrogen bonding network with the CO groups of Trp[3] and Tyr[5] and the NH groups of Tyr[5] and Leu[7]. The hydrophobic sidechains of Trp[3], Tyr[5] and Leu[7] which were once clustered together at the center of the turn, forming a "hydrophobic core" are again clustered together, this time above the turn, forming more of a "hydrophobic roof" over the buried Arg[8]. The potential energy of this conformation has dropped dramatically by 14.5 kcal/mole to ∼60kcal/mole from the higher energy minimum at 6 ps.

Eleven Picoseconds. After 11 ps, we observe a major rearrangement of the previous bend structure involving a "tightening" of the turn. The arginine has moved out of its position in the center of the bend, with a concomitant loss of the hydrogen bonds it was involved in, but despite this loss the resultant conformation is ∼1 kcal lower in potential energy than the structure after 8 ps. Part of the loss of the arginine interactions are made up by the formation of new hydrogen bonds between Trp[3] and Gly[10] with the Gly[10](CO) H-bonding to the Trp[3](NH) and Gly[10](NH) H-bonding to the Trp[3](CO) in a β-sheet type arrangement. In addition, the His[2] imidazole NH is hydrogen bonded to the Gly[10](CO), tying the amino and carboxy termini together. This conformation now contains a more "classical" bend, at the level of Gly[6]-Leu[7]. Although lacking the "characteristic" 4–1 hydrogen bond, the width of the molecule is roughly two amino acids instead of three as in the previous conformation.

COMPARISON OF GnRH CONFORMATIONS WITH THOSE OF A CYCLIC ANTAGONIST

The results summarized above, as well as a variety of experimental data,[28-32] show GnRH to be a highly flexible molecule. It is clear that alone, the knowledge of the accessible conformations of this hormone is not enough to gain an understanding of the conformational properties required for binding and transduction of the biological message it carries. More information is required. An approach to providing this information is to combine the data on the native agonist with other analogs making use of known sequence-activity relations. For example, the substitution of D-amino acids of glycine at position 6 (*i.e.,* D-Ala[6] or D-Trp[6] GnRH) results in enhanced biological activity, suggesting a β-turn conformation at the level of Gly[6]-Leu[7].[15] This is further supported by the lactam analog of Freidinger and co-workers.[33] These sequence-activity results would indicate that the turn conformations such as after 11 ps (FIGURE 1) are likely to be more significant with respect to biological activity than the extended structures.

Theoretical simulations may be used to further "quantitate" this type of reasoning by comparing the accessible structures of a series of analogs with the native hormone. The most desirable analogs to examine are cyclic peptides, where ring constraints are introduced by covalently linking residues well separated in the sequence. This severely limits the conformational freedom of the analog. Following this line of reasoning, we chose to consider the GnRH cyclic antagonist cyclo[Δ^3-Pro-D-Phe(p-Cl)-D-Trp-Ser-Tyr-D-Trp-NMe-Leu-Arg-Pro-β-Ala] (see below). In cultured pituitary cells, a 3.5-fold excess of the antagonist is required to reduce GnRH-induced secretion of LH by 50%, indicating that this analog binds to the GnRH receptor, although somewhat more weakly than GnRH itself.[7] By choosing an antagonist, we are limiting ourselves to consideration of the conformational features necessary for binding alone and making the first step toward the conformationally based design of GnRH antagonists.

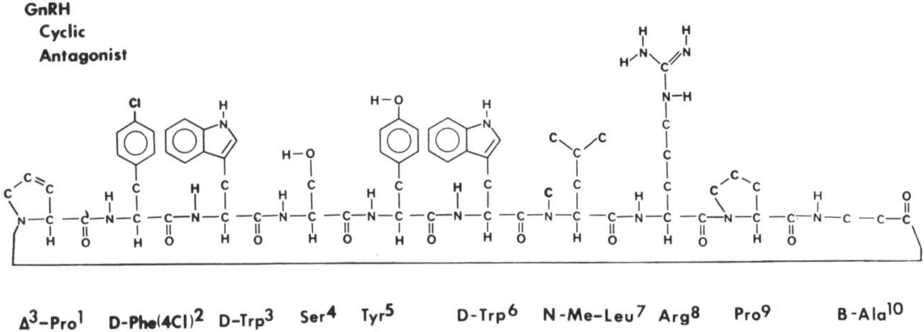

Conformations Accessible to the Cyclic Antagonist

The first step in studying the cyclic antagonist, as with GnRH itself, was to search the accessible conformational space of this molecule by molecular dynamics, followed by minimization of coordinates taken along the trajectory. The molecular dynamics trajectory spanned 24 ps and minimizations were carried out at 1 ps intervals along the trajectory. Several identical structures were found, leaving 18 distinct conformations. These 18 conformations constitute basically one family of conformations that share a common backbone conformation, but differ in the conformation of the sidechains.

This family is characterized by a hydrogen bonded turn at the level of Tyr^5-D-Trp^6-NMe-Leu^7-Arg^8, and a second chain reversal at residues β-Ala^{10}-$\Delta^{3,4}$-Pro^1-D-Phe^2-D-Trp^3 (FIGURE 2). The conformation of the ring is stabilized by three transannular hydrogen bonds: β-Ala^{10}(NH) . . . (CO)D-Trp^3, Tyr^5(NH) . . . (CO)Arg^8 and Arg^6(NH) . . . (CO)Tyr^5. In addition, there is a C$_7$ hydrogen bonded ring about $\Delta^{3,4}$-Pro^1 [D-Phe^2(NH) . . . (CO)β-Ala^{10}], and hydrogen bonds from the sidechains of Arg^8 and Ser^4 to mainchain carbonyls, both of which are also seen in GnRH conformations.

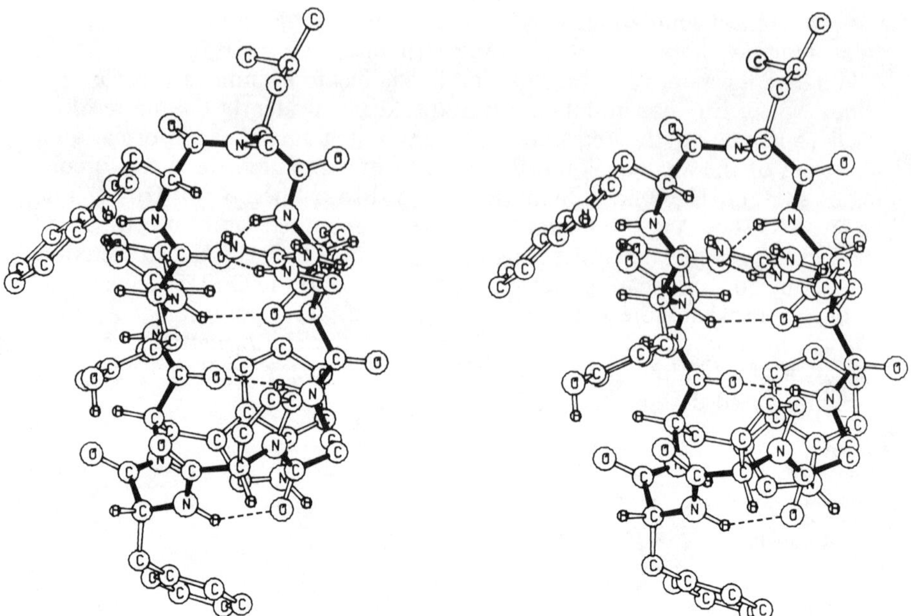

FIGURE 2. The minimum energy structure of the cyclic GnRH antagonist, cyclo[$\Delta^{3,4}$-Pro-D-Phe(4-Cl)-D-Trp-Ser-Tyr-D-Trp-NMe-Leu-Arg-Pro-β-Ala], derived from the instantaneous molecular dynamics configuration at 24 ps elapsed time during the trajectory.

Template Forcing

The preliminary results of the simulation of the dynamics of the cyclic analog indicate that only a limited set of conformations are accessible to this molecule. Furthermore, these conformations essentially constitute a single family characterized by only minor variations in the backbone structure corresponding to different sidechain orientations. We are now carrying out an extensive search to confirm these preliminary conclusions. Although several other high-energy conformations have been found, no low-energy conformations have been found. If there is only one conformational family accessible to the cyclic antagonist, then it follows that we have identified a putative "binding conformation." If this conformation indeed represents the "binding conformation," then it follows that portions of GnRH itself, which are common to both molecules, must be able to assume the conformation found in the cyclic antagonist. In order to test that this was indeed the case and the results are "internally consistent," we applied the technique of "template forcing," which we have recently developed, in order to answer the general question of how closely corresponding functional groups in two flexible molecules can be superposed.

Template forcing solves for the conformation of a molecule (in this case GnRH) that optimizes the overlap of functional groups with corresponding functional groups of a "template" conformation of another molecule (in this case, the GnRH cyclic antagonist), while at the same time maintaining the lowest cost in potential energy to the forced molecule. This is done by minimizing a function consisting of the sum of the expression for the potential energy of the forced molecule and a second term which is simply a constant times the root-mean-square distance between the common atoms in the two molecules:

$$F = V + K[\sum_{i}^{N}(x_i - x_i^0)^2/N]^{1/2} \tag{2}$$

In this equation, V is the potential energy of the system as expressed by the valence force field (1); x_i^0 are the coordinates of the atoms in the template corresponding to coordinates of the atoms of the forced molecule (x_i); N is the total number of atoms being forced; and K is a constant that controls the balance between potential energy and the degree of fit desired. Thus, the contribution from the forcing term decreases as the deviations between the template and the forced molecule decrease, and in the limit where the atoms are perfectly superposed the term vanishes.

We applied this technique to force GnRH to adopt a low-energy conformation of the cyclic antagonist. For this experiment, we applied the forcing to residues 4–8 and 4–9, with the exception of the sidechain of D-Trp[6] in the antagonist (Gly[6] in GnRH) and the methyl group on the nitrogen of NMe-Leu[7] (Leu[7] in GnRH). The results of this "experiment" indicate that very close fits of 0.29 Å for residues 4–8, and 0.38 Å for residues 4–9, can be achieved at a low energy cost (of ~5 kcal/mole relative to the lowest energy conformation of GnRH found so far). Given this close fit between the two analogs and the low cost in energy required to achieve it (*i.e.,* which would easily be induced on binding to the receptor[34]) the conformation of the cyclic antagonist is consistent with a putative binding conformation, at least for the region contained in residues 4–9. It therefore can be used productively as a basis for further conformationally based design, in an iterative procedure, either to confirm this conformational hypothesis or to elucidate fallacies in the model.

An Example of Conformationally Based Analog Design

Having identified a putative binding conformation for the cyclic antagonist, we desired to test the validity of that conformation by introducing additional covalent constraints designed to "lock in" unambiguously the proposed conformation. A common feature found in the low-energy conformations of the cyclic antagonist is that the α-hydrogens of Ser[4] and Pro[9] are pointing into the center of the ring and roughly across from one another in the β-structure connecting the two β-turns (FIGURE 2). From this structure, it would appear that the α-carbon atoms could be bridged by either two or three methylene groups (-CH$_2$-CH$_2$ or -CH$_2$-CH$_2$-CH$_2$-) without in-

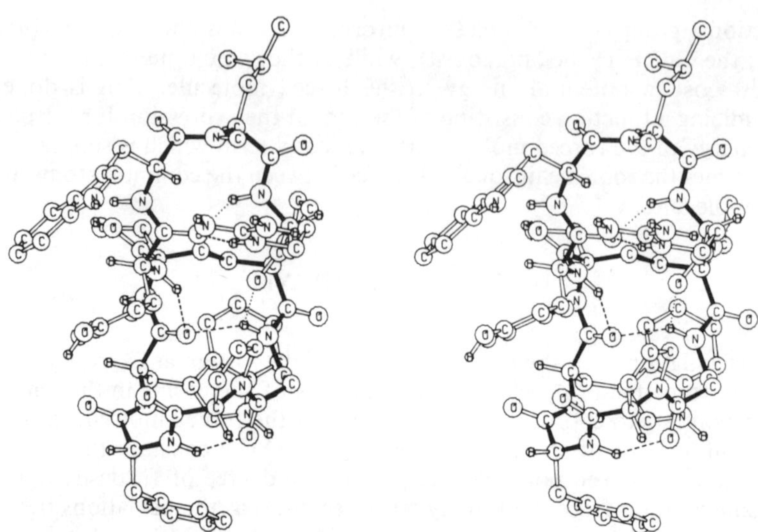

FIGURE 3. The minimum energy structure of the proposed transannular bridged analog where the α-carbons of Ser[4] and Pro[9] have been linked by an ethylene bridge to constrain the conformation to that found for the parent cyclic antagonist.

troducing large perturbations into the rest of the molecule. Such bridging groups would provide analogs where the two strands of β-sheet are covalently locked into proximity. In order to check that the proposed bridges did not disrupt the structure in other respects, we built the two analogs in the computer and minimized their energy with respect to all degrees of freedom. The resulting structure for the analog bridged by two methylenes is shown in FIGURE 3. Both analogs minimized to conformations very near the parent cyclic antagonist conformation (0.3 Å in the CH_2-CH_2 bridge and 0.6 Å for the CH_2-CH_2-CH_2 bridge). We are presently investigating possible synthetic routes to this analog and are in the process of designing more readily synthesizable analogs with similar objectives in mind.

The spectroscopic analysis of both the parent and the bridged analog should result in essentially equivalent spectra, indicating that the parent cyclic antagonist exists in a conformation that is within the even more limited set of conformations available to the bridged analog. The bridged analog, by the nature of its covalent architecture, is constrained to conformations similar to those proposed for the present cyclic antagonist. Therefore, bridged analogs of this type can be used to test the validity of the conformation proposed for the parent cyclic antagonist. Further, if the bridged analog itself were to prove active, then this would be very strong evidence in favor of the putative "binding conformation."

REFERENCES

1. RIVIER, C., W. VALE & J. RIVIER. 1983. J. Med. Chem. **26**: 1545.
2. HSUEH, A. J. W. & P. B. C. JONES. 1983. Ann. Rev. Physiol. **45**: 83.
3. MATSUO, H., Y. BABA, R. M. G. NAIR, A. ARIMURA & A. V. SCHALLY. 1971. Biochem. Biophys. Res. Comm. **43**: 1334.
4. BURGUS, R., M. BUTCHER, M. AMOSS, N. LING, M. MONAHAN, J. RIVIER, R. FELLOWS, R. BLACKWELL, W. VALE & R. GUILLEMIN. 1972. Proc. Natl. Acad. Sci. U.S.A. **69**: 278.
5. ZATUCHNI, G. I., J. D. SHELTON & J. J. SCIARRA, Eds. 1981. LHRH Peptides as Male and Female Contraceptives. Harper & Row. Philadelphia.
6. RIVIER, J., M. BROWN, C. RIVIER, N. LING & W. VALE. 1976. *In* Peptides 1976. Proceedings of the Fourteenth European Peptide Symposium. A. Loffet, Ed: 427. Editions de l'Université de Bruxelles. Brussels.
7. RIVIER, J., C. RIVIER, M. PERRIN, J. PORTER & W. VALE. 1981. *In* LHRH Peptides as Male and Female Contraceptives. G. I. Zatuchni, J. D. Shelton & J. J. Sciarra, Eds.: 13. Harper & Row. Philadelphia.
8. VALE, W., G. GRANT, J. RIVIER, M. MONAHAN, M. AMOSS, R. BLACKWELL, R. BURGUS & R. GUILLEMIN. 1972. Science **176**: 933.
9. PERRIN, M. H., J. E. RIVIER & W. W. VALE. 1980. Endocrinology **106**: 1289.
10. REES, R. W. A., T. J. FOELL, S. CHAI & N. GRANT. 1974. J. Med. Chem. **17**: 1016.
11. RIVIER, J., N. LING, M. MONAHAN, C. RIVIER, M. BROWN & W. VALE. 1975. *In* Peptides: Chemistry, Structure and Biology. Proceedings of the Fourth American Peptide Symposium. R. Walter & J. Meinhofer, Eds.: 863. Ann Arbor Science Publishers. Ann Arbor.
12. RIVIER, J. & W. VALE. 1978. Life Sci. **23**: 869.
13. NEKOLA, M. V., A. HORVATH, L. J. GE, D. H. COY & A. V. SCHALLY. 1982. Science **218**: 160.
14. NESTOR, J. J., R. TAHILRAMANI, T. L. HO, G. I. MCRAE, B. H. VICKERY & W. J. BREMNER. 1983. *In* Peptides: Structure and Function. Proceedings of the Eighth American Peptide Symposium. V. H. Hruby & D. H. Rich, Eds.: 861. Pierce Chemical Co. Rockford, Ill.
15. MONAHAN, M., M. AMOSS, H. ANDERSON & W. VALE. 1973. Biochemistry **12**: 4616.
16. HAGLER, A. T. *In* Physical Methods in Peptide Conformational Studies. The Peptides, vol. 7. In press.
17. HAGLER, A. T., P. S. STERN, R. SHARON, J. BECKER & F. NAIDER. 1979. J. Am. Chem. Soc. **101**: 6842.
18. DAUBER, P. & A. T. HAGLER. 1980. Accts. Chem. Res. **13**: 105.
19. LIFSON, S., A. T. HAGLER & P. DAUBER. 1979. J. Am Chem Soc. **101**: 5111.
20. HAGLER, A. T., S. LIFSON & P. DAUBER. 1971. J. Am. Chem. Soc. **101**: 5122.
21. HAGLER, A. T., P. DAUBER & S. LIFSON. 1979. J. Am. Chem. Soc. **101**: 5131.
22. HAGLER, A. T., P. S. STERN, S. LIFSON & S. ARIEL. 1979. J. Am. Chem. Soc. **101**: 813.
23. HAGLER, A. T. & A. LAPICCIRELLA. 1976. Biopolymers **15**: 1167.
24. HAGLER, A. T., E. HULER & S. LIFSON. 1974. J. Am. Chem. Soc. **96**: 5319.
25. HAGLER, A. T. & S. LIFSON. 1974. J. Am. Chem. Soc. **96**: 5327.
26. LAU, S. H., J. RIVIER, W. VALE, E. T. KAISER & F. J. KEZDY. 1983. Proc. Natl. Acad. Sci. U.S.A. **80**: 770.
27. SPRECHER, R. F. & F. A. MOMANY. 1979. Biochem. Biophys. Res. Comm. **87**: 72.

28. KOPPLE, K. D. 1981. *In* Peptides: Synthesis-Structure-Function. Proceedings of the 7th American Peptide Symposium. D. H. Rich & E. Gross, Eds.: 295. Pierce Chemical Co. Rockford, Ill.
29. CANN, J. R., K. CHANNABASAVAIAH & J. M. STEWART. 1979. Biochemistry **18**: 5776.
30. DESLAURIERS, R., G. C. LEVY, W. H. McGREGOR, D. SARANTAKIS & I. C. P. SMITH. 1975. Biochemistry **14**: 4335.
31. WESSELS, P. L., J. FENNEY, H. GREGORY & J. J. GORMLEY. 1973. J. Chem. Soc. Perkin Trans. **2**: 1691.
32. DONZEL, B., J. RIVIER & M. GOODMAN. 1977. Biochemistry **16**: 2611.
33. FREIDINGER, R. M., D. F. VEBER, D. S. PERLOW, J. R. BROOKS & R. SAPERSTEIN. 1980. Science **210**: 656.
34. JENCKS, W. P. 1975. Adv. Enzymol. Relat. Areas Mol. Biol. **43**: 219.

Engineering Aspects of Protein Structure[a]

F. R. SALEMME

Genex Corporation
Gaithersburg, Maryland 20877

X-ray crystallographic studies have revealed proteins to be highly organized structures of marvelous complexity. A protein's three-dimensional organization is required for biological activity because it provides the spatial framework necessary for accurate positioning of amino acid residues involved in ligand binding or substrate catalysis. Nevertheless, many studies indicate that the functional arrangements manifest in crystal structures have only marginal stability in solution.[1] As a result, proteins exhibit a wide range of dynamical behavior at ambient temperatures.[2-5] Although dynamical excitation of a protein may play a role in catalysis, the continual statistical interchange of kinetic energy between a protein and its solvent environment can lead to transient or irreversible disruptions of structure and function. The basic question addressed here concerns how the long-range structural integrity of a protein is preserved in a dynamically active solution environment.

Information relevant to this question comes from comparative studies of the large number of protein structures presently known from X-ray crystallography.[6] These studies have shown that proteins can be structurally classified and that features of extended structural organization recur among proteins having no apparent sequence or evolutionary relationship. The recurrence of these structural patterns must then reflect an accomodation to some basic physical requirements of facile folding or stability, for which there apparently exist a finite number of structural solutions.[7-8] What we wish to illustrate here is how some recurrent long-range architectural features of proteins achieve and maintain their structural integrity in a kinetically active environment.

It has long been recognized that extended features of protein tertiary organization are generally associated with the formation of hydrogen-bonded secondary structures. Owing to the periodic and structurally invariant nature of the polypeptide backbone, long regions of chain can form regular lattices stabilized by hydrogen-bonded interactions. These extended lattice

[a] This work was supported by research grant nos. GM 30393 and GM 33325 from the National Institutes of Health.

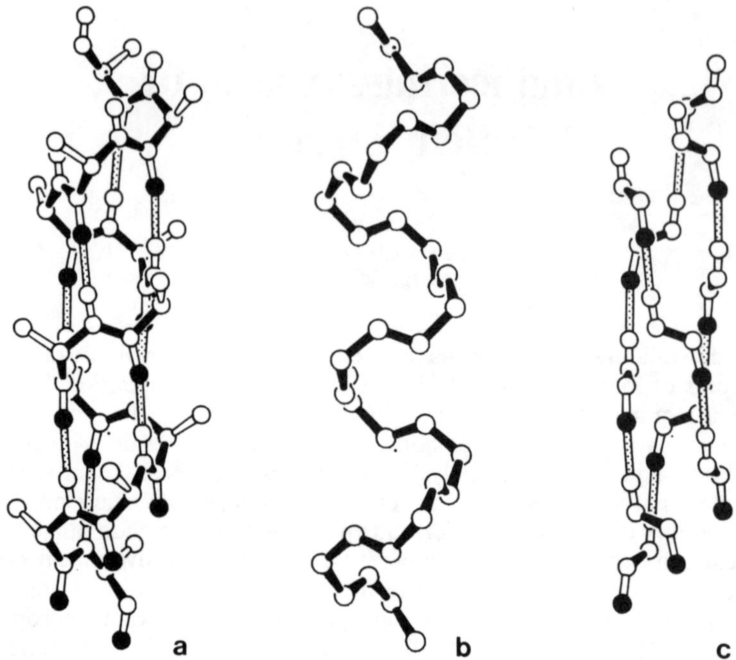

FIGURE 1. A molecular model of an α-helix (**a**). As illustrated in (**b**) and (**c**), the structure is organized as a cylindrical lattice formed of a right-handed backbone helix integrally braced by three left-handed hydrogen-bonded helices.

structures provide the basis for long-range structural organization in the protein.

FIGURE 1a illustrates an α-helix, one of the most common extended secondary structural arrangements seen in proteins. Theoretical studies of α-helix dynamics are consistent with structural observations indicating that this is a particularly rigid arrangement that tends to resist deformation.[9] The mechanical origins of this ridgidity are readily understood if the backbone covalent and hydrogen-bonded interactions are separated as illustrated in FIGURES 1b and 1c. These make evident that atoms of the sequentially connected backbone form a right-handed helix with 3.6 residues per turn, whereas the contiguous hydrogen-bonded interactions form a triple-strand left-handed helical arrangement. Deformation of any helix by stretching or bending will result in the helix unwinding in a direction opposite to the original helix twist. In the case of the α-helix, the minimum energy structure is one that simultaneously optimizes nonbonded interactions in the right-hand helix, and hydrogen-bonded interactions in the left-hand helices. Deformations of the extended α-helix are strongly resisted because these must involve the simultaneous unwinding of helices of opposite hand. This is physically impossible since the helical interactions in the α-helix are intergrally connected to form a continuous cylindrical lattice. The objective of treating the α-helix from this per-

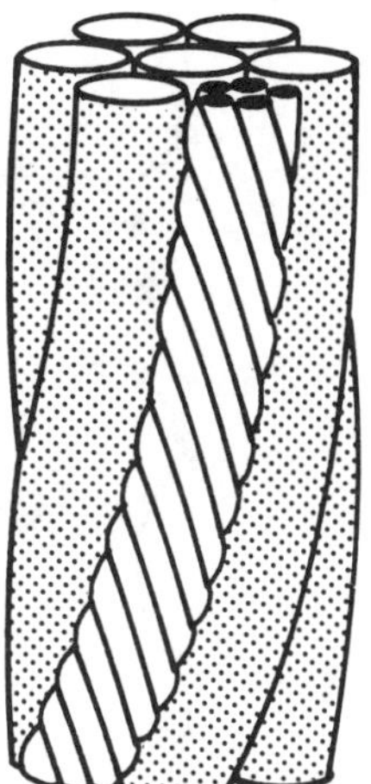

FIGURE 2. A schematic view of a wire cable. Resistance to stretching and twisting is achieved by laying the strands in each wire with opposite twist sense to the lay of the wires in the cable.

spective is to emphasize the distributed nature of the interaction forces that stabilize the extended structure. The basic principle of utilizing the opposition in twist sense to resist extension or disassembly in helical structures has, in fact, been exploited for hundreds of years in the manufacture of rope and cables (FIG. 2).

A second common element of protein secondary structure is the parallel β-sheet. As first described by Pauling and Corey,[10] the classical structure is composed of extended two-fold helical chains arranged with the same N to C polypeptide chain sense. This allows the formation of a regular hydrogen-bonding pattern between chains so that the structure as a whole forms a planar lattice connected by covalent bonds along the chains' direction and hydrogen bonds across the strands of the sheet (FIG. 3). The structure determination of numerous proteins showed that in contrast to the ideal flat model, observed sheets in proteins virtually always formed complex twisted surfaces (FIGS. 4–6), and that the sense of the twist in the sheet was always the same.[11] Subsequent analysis[8] of these structures showed that the overall sheet twist resulted from twisting of its constituent chains from the two-fold helical conformation of the flat sheet, into left-hand helices with 2.1 to 2.5 residues per turn. The polypeptide chain twisting is a consequence of the structures' incorporation of L-amino acids, which energetically prefer left-twisted conformations in extended chains. As the flat structure is one that optimizes interchain hydrogen bonds, it is evident that twisting the sheet to optimize the local chain conformational energy can only occur at the expense of the interchain hydrogen-bond energies. The lattice formed by a twisted sheet is consequently stressed, with the components of conformational energy that drive chain twist acting in opposition to the optimized arrangement of the interchain hydrogen bonds.[8] It is the distributed character of the periodic lattice interactions that give rise to the remarkably uniform curvatures in the observed structures, whereas it is the opposition of helices twisting with op-

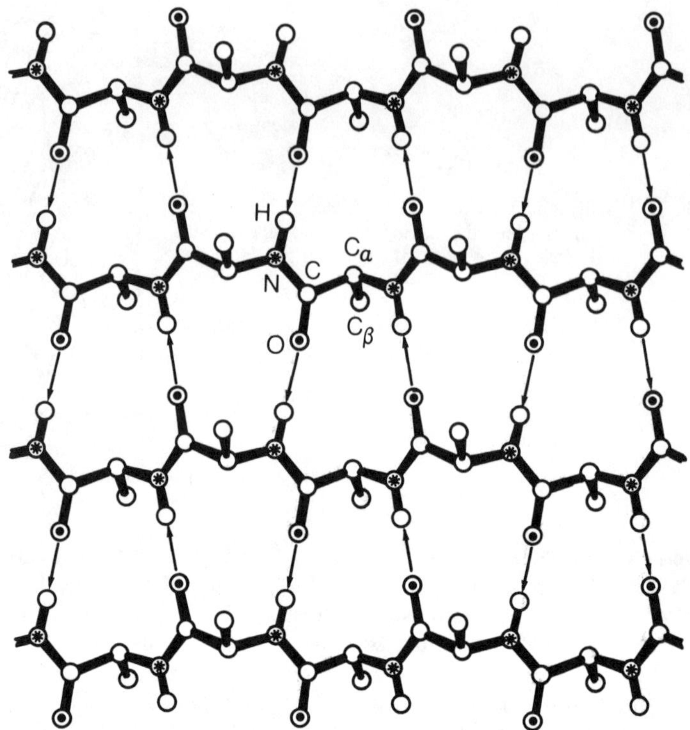

FIGURE 3. A schematic view of a flat, multiple-strand parallel β-sheet. The structure forms a lattice stabilized by covalent interactions along the chain directions and hydrogen-bonded interactions across the chains.

posite senses that results in a rigidly defined extended geometry for the structure as a whole (FIG. 7).

Stressed structures incorporating both the anticlastic curvature and geometrical properties inherent in protein parallel sheets have found wide architectural application in the construction of light and rigid structures such as roofs and power plant cooling towers.[12] Consequently, it is perhaps not surprising to find that nature has used the same engineering principles in the construction of extended and rigid backbone structures in proteins.

The preceding has emphasized how chiral forces ultimately arising from the polypeptide chains' composition of L-amino acids manifest themselves as operative factors stabilizing extended structures in proteins. In the representative examples described above, the opposition of twist components was involved in the maintenance of the extended structural integrity required for the preservation of protein function in a dynamically active environment.

Many additional invariant chiral effects are observed in known proteins

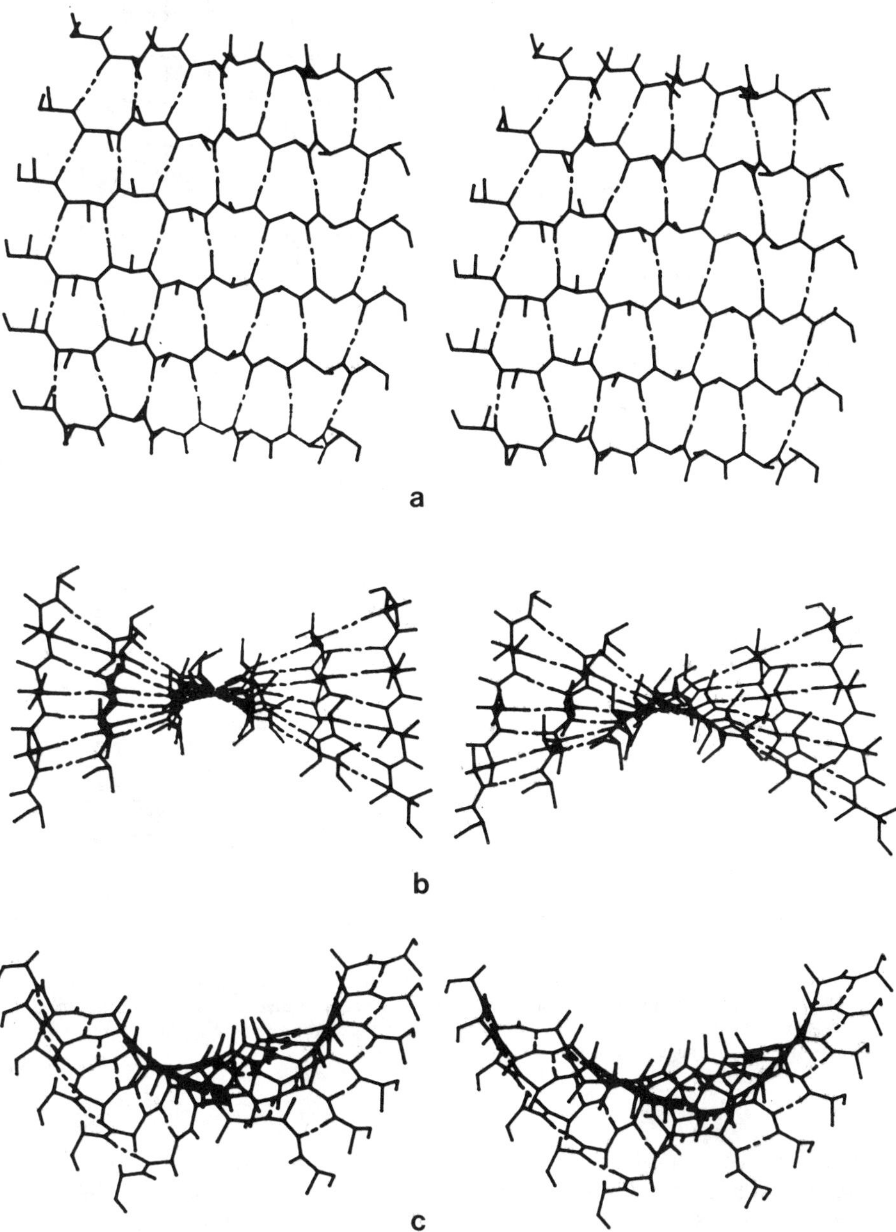

FIGURE 4. Stereo views of a rectangular plan, twisted parallel β-sheet.

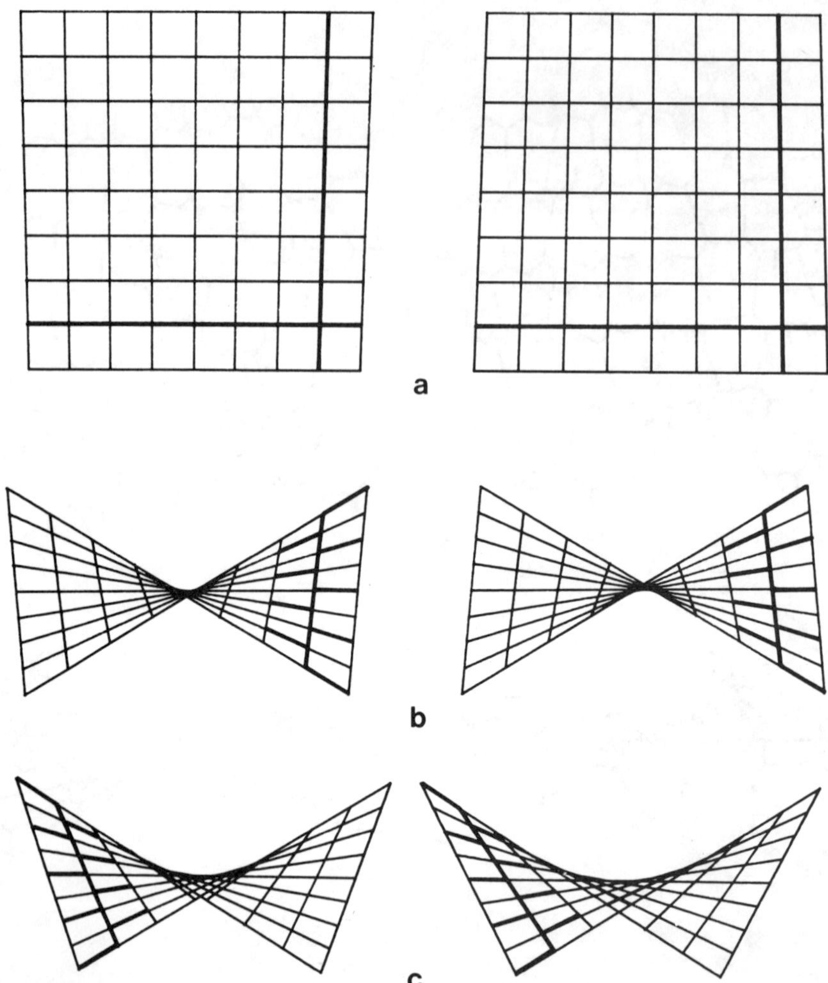

FIGURE 5. Schematic views of the hyperbolic parabolid surface formed by parallel sheets as shown in FIGURE 4. Note that this uniformly curved surface results from stress equilibration among two sets of helical interactions lying respectively along the polypeptide chains and along the contiguous hydrogen bond networks between strands.

involving, for example, the handedness of looping interconnections between stretches of polypeptide chain forming extended secondary structures.[6] Generally these loops form right-handed supercoils, which again may reflect a statistical preference for forming locally left-handed conformations in extended polypeptide chains. Additional manifestations of chiral invariance are present in extended antiparallel β-sheets. However, in this case, the prop-

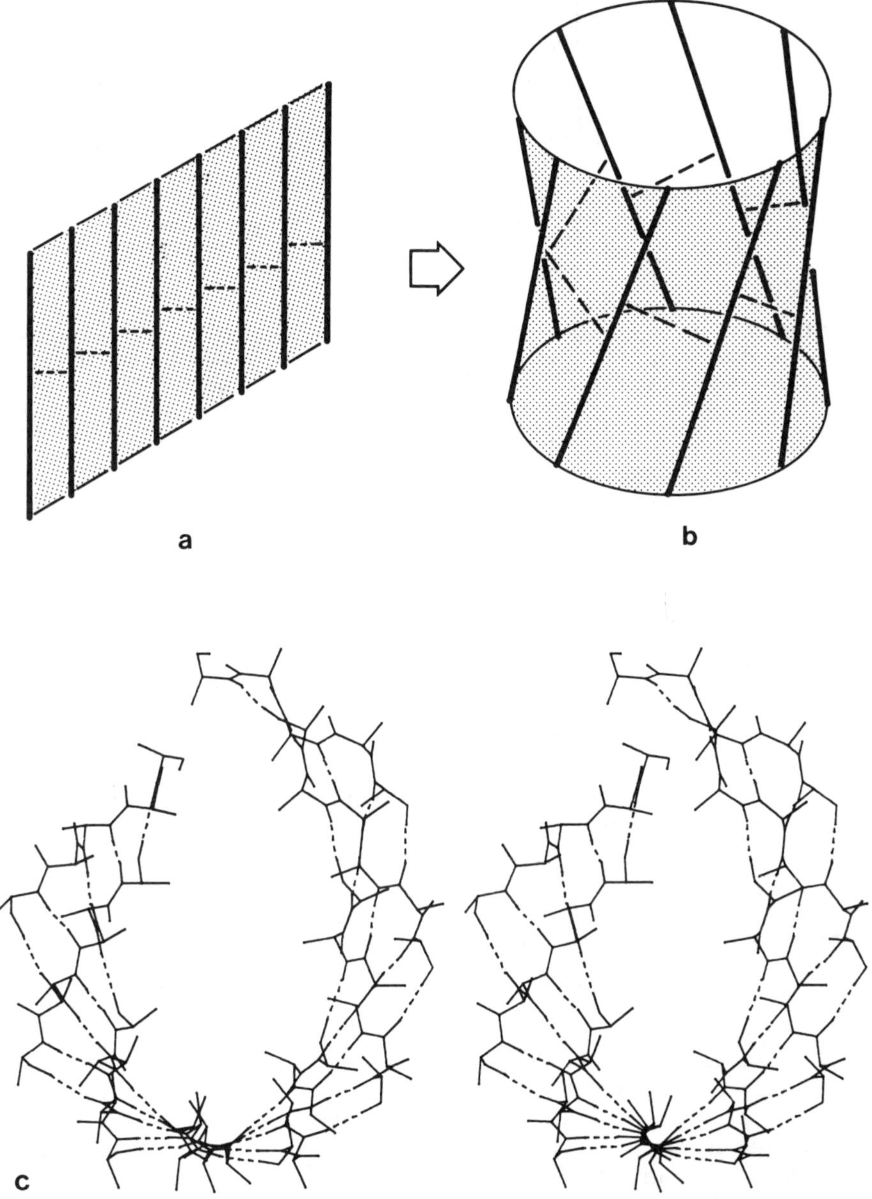

FIGURE 6. (a, b) The formation of parallel sheet structures with cylindrical curvature results from the pattern of hydrogen-bonded interactions between adjacent chains. (c) A stereo view down a helical chain axis of a cylindrically curved sheet with a staggered-chain hydrogen bonding pattern. Both the structures shown in FIGURES 5 and 6b are locally curved in opposite directions (anticlastically curved), an arrangement that makes such stressed lattice structures particularly rigid.

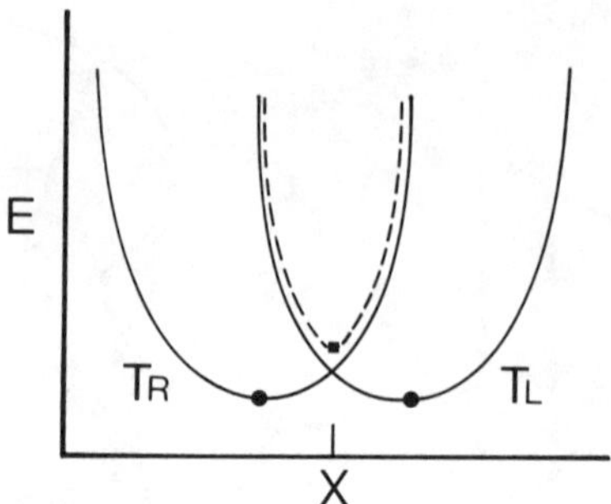

FIGURE 7. A plot of potential energy versus displacement in extended stressed structures. The structural stabilization of extended lattice structures can be usefully viewed as a result of the interaction of potential surfaces that reflect delocalized twist components in the structure as a whole (*solid curves*). The stressed configuration is constrained by the potential (*dashed*) defined by the opposing twist components. Although the individual twist potentials may be relatively soft (*e.g.*, small changes in potential energy occur for relatively large variations in displacement), the dashed potential produced by twist forces acting in opposition results in larger variations in energy with displacement, and consequently a more rigid structural configuration.

erties observed suggest that long-range cooperative interactions associated with sheet hydrogen-bond formation play a role in structure folding.

Multiple-strand antiparallel sheet domains virtually always appear to have been assembled by a folding pathway that first involves the formation of a long double-strand "hairpin" sheet. Subsequent assembly then appears to involve the concerted right-handed supercoiling of the double-strand sheet to produce the final structural domain (FIG. 8). This hypothetical mechanism[6] for folding sheets is of interest in the present context because it infers that the hairpin sheet is a highly flexible structure with a built-in tendency to form right-handed supercoils. Detailed studies of the conformational flexibility of double-strand antiparallel sheets indeed show them to possess an unusual extent of cooperative flexibility.[8] Most interestingly, the structure can be shown to supercoil perferentially with a right-handed sense to produce structures that are topologically equivalent to the observed domains in proteins (FIG. 9). These results indicate that the periodic hydrogen bonding constraints shown above to be important in the protein's final structural organization may, in the case of antiparallel sheet domains, additionally play a determinative role in defining a highly cooperative domain folding pathway. In a mechanical sense, the cooperative supercoiling of the double-strand hairpin sheet arises from its kinematic properties, *i.e.*, how the structure undergoes characteristic cooperative motions as defined by the nature and geometry of its mechanical linkages. Alternatively, this kinematic behavior can be viewed as the result of the superposition of delocalized vibrational modes of the structure as a whole.[5] In this case, initial formation of the hydrogen-bonded interactions in the double-strand sheet serve to couple the structure, thus resulting in the partitioning of the structure's kinetic energy into a new set of vibrational modes characteristic of the structure as a whole.[8] It is the su-

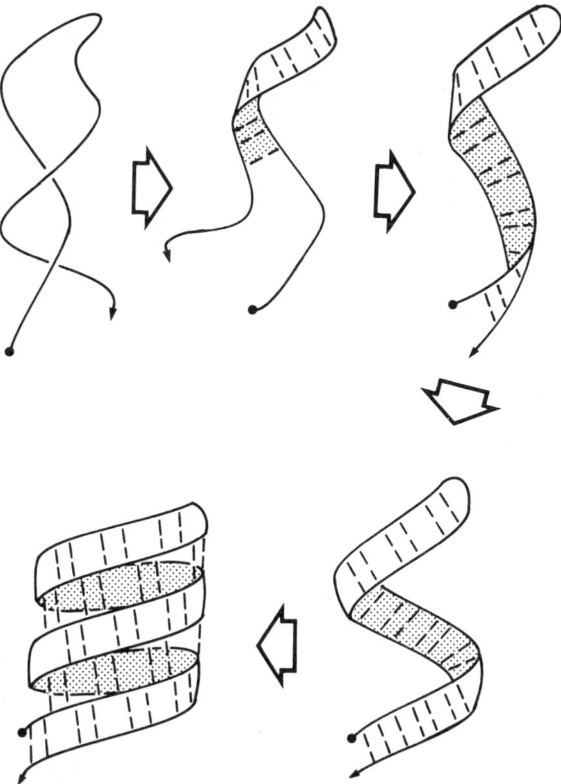

FIGURE 8. A schematic illustration of a hypothetical folding pathway for an antiparallel β-roll domain as observed in some proteins of known structure (modified after an illustration by Jane Richardson). Structure folding proceeds by the right-hand supercoiling of an initially formed double-strand, "hairpin" antiparallel sheet. The chiral folding trajectory appears to be determined by the cooperative folding properties of the double-strand structure (FIGURE 9).

perposition of these low-energy modes that reflect themselves in the structure's kinematic properties and apparently must be involved in the process of folding β-sheet domains.

We have shown that many aspects of long-range order in proteins reflect the distribution of forces over extended lattice structures. In the case of α-helices or parallel β-structures, these distributed forces effectively serve to prestress the structure, so conferring rigidity in a dynamically active environment. Although extended lattice interactions play similar roles in the organization of antiparallel sheet domains,[8] here they additionally play a determinative role in the process of protein folding. Although many questions remain concerning the origins of protein structural stability, we have illustrated here that many of their properties plainly reflect adherence to well-known principles of structural engineering.

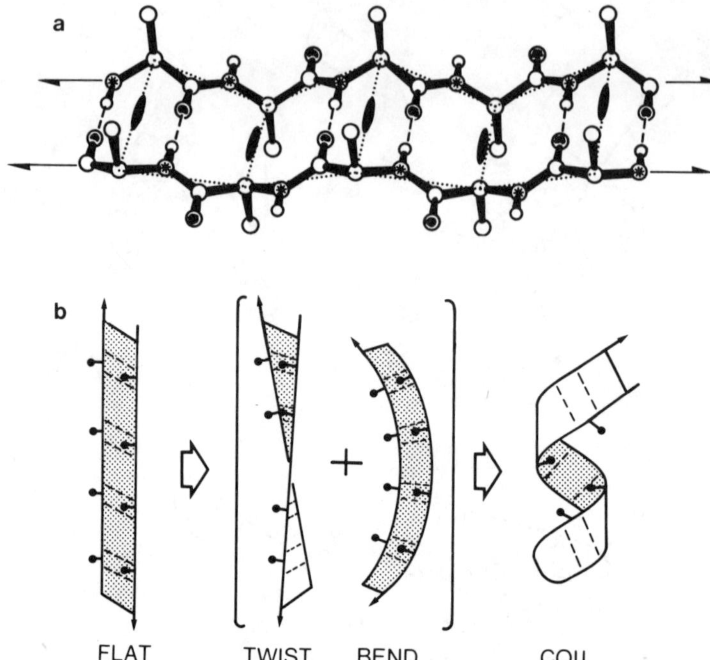

FIGURE 9. A section of double-strand antiparallel sheet illustrating that opposite surfaces of the sheet differ (**a**); *e.g.,* one side is populated by substituents from 10-membered hydrogen-bonded rings, and the other by substituents from 14-membered rings. Analysis of the low energy deformation pathways for the flat sheet show that the observed right-supercoiling results from coupled superposition of twist and bend components (**b**). Both twist and bend have an energetically preferred sense, so that resulting double-strand coils always have small ring substituents located on the coil interior. This feature is also preserved in the observed structures illustrated in FIGURE 8 (*shaded surfaces*), and so provides primary evidence of the role of cooperative double-strand coiling as a pathway for their assembly.

REFERENCES

1. PRIVALOV, P. L. 1979. Adv. Prot. Chem. **33**: 167–241.
2. CARERI, G., P. FASELLA & E. GRATTON. 1979. A. Rev. Biophys. Bioeng. **8**: 6997.
3. MCCAMMON, J. A., B. R. GELIN & M. KARPLUS. 1977. Nature **277**: 585–590.
4. COOPER, A. 1976. Proc. Nat. Acad. Sci. U.S.A. **73**: 2740–2741.
5. PETICOLAS, W. L. 1979. Meth. Enzymol. **61**: 425–456 (part H).
6. RICHARDSON, J. S. 1981. Adv. Prot. Chem. **34**: 167–339.
7. WEBER, P. C. & F. R. SALEMME. 1980. Nature (London) **287**: 82–84.
8. SALEMME, F. R. 1983. Prog. Biophys. Molec. Biol. **42**: 95–133.
9. LEVY, R. M. & M. KARPLUS. 1979. Bipolymers **18**: 2465–2995.
10. PAULING, L. & R. B. COREY. 1951. Proc. Nat. Acad. Sci. U.S.A. **37**: 729–740.
11. CHOTHIA, C. 1973. J. Molec. Biol. **75**: 295–302.
12. OTTO, F. 1969. Tensile Structures. MIT Press. Cambridge, Mass.

Dynamic Aspects of
Protein Structure

MARTIN KARPLUS

Department of Chemistry
Harvard University
Cambridge, Massachusetts 02138

Before I turn to my primary topic, the dynamics of macromolecules, I shall make a few remarks that may be useful in the present context. The primary objective of the work reported in this volume is to learn how to "look" at protein structures in more detail than one has been able to in the past. Two aims of such "looking" are to predict the fold of proteins of unknown structure from the known structures of homologous proteins and to understand the binding of ligands, substrates and inhibitors to proteins. What is illustrated in this volume is that we are, in fact, learning to look more effectively. We are doing this with molecular graphics and with energy functions, two fields in which great strides have been made in recent years. There may be a tendency, with the availability of such seductive pictures and complicated programs that seem to be able to do everything, to forget about the limitations of the methodology and to follow blindly what the pictures or energy calculations seem to be telling you.

To emphasize this point I should like to recall how this field originated before the present technology came into being. Crystallographers have looked at the X-ray structures of proteins for a long time and they are still looking at these structures, whether as models in space or as models on a computer graphics display. A large amount of information has been deduced in this way about what goes on in these rather complex systems. The new graphics technologies obviously make it much easier to look, and the energy calculations provide an added dimension to such "looking." When examining two closely related structures, one tries to determine from the nature of the differences between the two how to go from one structure to another. In so doing one thinks about the effect of nonbonded contacts, hydrogen bonds and related interactions. What the energy functions do is make possible a somewhat more quantitative evaluation of such interactions. Ramachandran-type hardsphere models are replaced by soft Lennard-Jones interactions. Perhaps more important, complicated longer range effects that would be difficult to find by looking can be made evident by the calculations; *e.g.,* when atom a is displaced, whether atom b, c or d moves the most may depend on complex connectivities that are included in the energy function, or longer range electrostatic interactions that are difficult to visualize from a model.

There are many worthwhile things that one can do with molecular graphics and energy minimization techniques, but it is necessary to remember the fact that they are only aids to our thinking; if one does not think, nothing useful will come out of the methods. Since I have never done any modeling of the type that is the main theme of the conference, it is easy for me to make such cautionary comments, but they are important to remember, nevertheless.

THE QUESTION OF ELECTROSTATIC INTERACTIONS

As an example of the need to be careful about the use of empirical energy functions, we consider the role of electrostatic interactions in determining certain properties of a simple model system, N-methyl alanyl acetamide ($CH_3CONHCHCH_3CONHCH_3$), often referred to as the alanine "dipeptide." For this study, carried out by Dr. B. M. Pettitt,[1] a standard empirical potential of the type already described by Dr. A. Hagler (this volume, see also ref. 2) was used. Since the electrostatic interactions in most empirical energy functions are introduced by placing point charges on the atoms, the values of the charges were varied while keeping all other parameters constant.

We consider here two extreme sets of charges; one set has zero charges on the atoms and the other corresponds to that used in the CHARMM program.[2] The latter has (0.48, -0.48) for a C-O group, (0.36, -0.26) for a NH group, (-0.3, 0.1, 0.1, 0.1) for a CH_3 group and (0.1, 0.1) for $C^\alpha H$, all in units of the electronic charge.

Given these two sets of charges, it is of interest to ask what effect they have on the structure, dynamics and thermodynamics of the dipeptide. The thermodynamics is expected to be dominated by the energy itself, although the vibrational entropy also can play a role, particularly in the relative free energies of different conformers. As to the structures of the conformers as a function of the dihedral angles ϕ and ψ, they are determined by the first derivatives of the energy. Finally, the motion in the neighborhood of each conformational minimum is a function of the second derivatives of the energy, from which are obtained the force constants for a vibrational analysis and an evaluation of the harmonic dynamics. Since the electrostatic interactions depend on $1/r_{ij}$, where r_{ij} is the distance between particles i and j, they are slowly varying functions of distance in comparison with the Lennard-Jones interactions. Thus, one would expect them to have a large effect on the absolute values of the calculated energies, but a much smaller effect on the structure and dynamics, which depend on energy derivatives.

As far as the total energy is concerned, changing the charges makes a large difference, up to 10 kcal/mole for this small molecule. As can be seen from TABLE 1, the relative energies of the various minima change less, although still by up to 3 kcal/mole. When the structures of the minima are compared, it is evident that the changes as a function of charge are rather small; the largest effect occurs for the angle ψ ($\sim 4°$). The fifteen lowest normal mode frequencies that make the dominant contributions to the atomic displacements are listed in TABLE 2. The lowest mode, which is composed

TABLE 1. Major Minima on the ϕ, ψ Surface[a]

	ϕ	ψ	E^b	ΔE^c
Zero charges				
C_{7eq}	-69.2	65.1	15.6	0.0
C_{7ax}	64.9	-63.5	15.7	0.1
C_s	-174.5	174.3	17.4	1.8
CHARMM charges				
C_{7eq}	-67.9	61.3	4.9	0.0
C_{7ax}	64.0	-59.6	5.2	0.3
C_s	-172.8	176.6	9.7	4.8

[a] All angles in degrees and energies in kcal/mole.
[b] Total energy (relative to arbitrary zero).
[c] Energy relative to absolute minimum for a given model.

mainly of the ϕ, ψ fluctuations, changes by about 30%; the higher modes change by less than 10%. The room temperature root-mean-square fluctuations and vibrational contributions to the enthalpy and entropy, all of which involve sums over the modes, are altered negligibly.

These results suggest that the motional properties are relatively insensitive to the electrostatic interactions, but that for total energies accurate electrostatic potentials are essential. Unfortunately, it is quantities corresponding to such total energies that play the important role in the binding of an inhibitor or a substrate.

THE REALITY OF X-RAY STRUCTURES

In addition to cautions about theoretical analyses, it is useful to introduce cautions concerning experimental results. As Richard Feldmann has empha-

TABLE 2. Lowest Normal Mode Frequencies[a]

Number	Zero charges	CHARMM charges
1	41.0	59.2
2	71.3	71.6
3	116.7	128.3
4	127.6	138.0
5	162.8	157.7
6	179.0	178.2
7	218.4	222.1
8	233.9	239.1
9	253.1	258.8
10	300.0	301.5
11	324.0	337.2
12	363.3	360.1
13	440.6	439.3
14	471.3	476.4
15	554.9	542.9

[a] Values are in wave numbers.

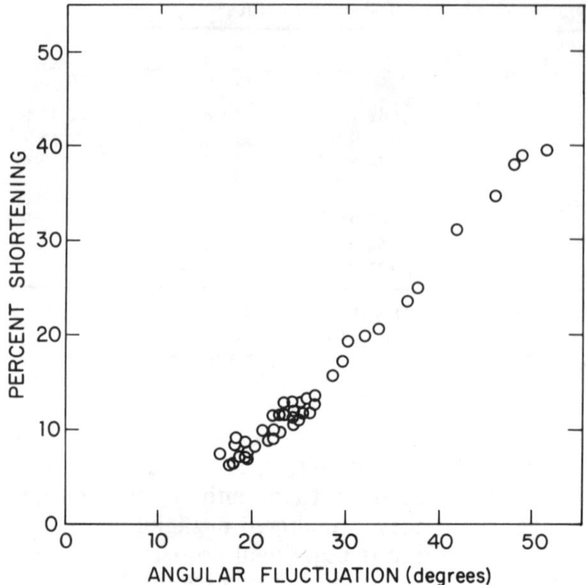

FIGURE 1. Bond shortening in average structure (see text); each circle represents an individual bond.

sized (this volume), X-ray crystallography is the basis of much of the modeling work that is being done for macromolecules. It is important to remember, however, that the structure obtained from an X-ray analysis does not exist. X-ray crystallography at best finds an average structure—it is an average over the large number of molecules in the crystal (disorder) and over a long period of time (dynamics). The probability that a given macromolecule actually has the average structure at any time is nearly zero. Nevertheless, it is clear that if the fluctuations are not too large and the disorder is not too great, the structure is an interesting one and can be used, as it has been, to obtain rather detailed information.

Recently, we have begun to use dynamical simulations to examine whether significant errors can be introduced into the structure by the assumptions made in the refinement procedure. To illustrate this point, I shall refer to some work by H. Yu[3] concerning the effect of dynamics on the observed bond lengths. Since the X-ray results correspond to the average structure, we can simulate the expected results by doing a molecular dynamics run, finding the average coordinates from that run, and calculating bond lengths or other structural parameters from the average coordinates. This is to be contrasted with the "real" average bond lengths, which are evaluated by determining the bond lengths for each individual coordinate set and then averaging the calculated bond lengths over the trajectory. Yu used a simulation of the bovine pancreatic trypsin inhibitor (BPTI),[4] and obtained the results shown in FIGURE 1 for the bonded pairs of atoms. The abscissa represents the root-mean-square angular fluctuation of a bond and the ordinate its percent shortening in the average structure, relative to the average bond length. There are 468 bonds in the extended atom model of BPTI; the bonds that have an effective shortening of more than 6% are shown in the figure. It can be seen that there are about 30 bonds that are contracted by more than 10% and a few that are contracted by 40%; the latter, for a C-C bond of 1.5 Å, would correspond to an observed bond length of 0.9 Å. Although such large contractions are rare, the effect is nevertheless significant. It indicates that refining with constraints that keep the bond lengths in the average structure close to a standard value can introduce non-negligible errors.

This effect is well known in small molecule crystallography,[5] but it is likely to be more important in proteins where the fluctuations are larger. What is happening is that the bond is taking on different orientations in space because of the motion of the bonding atoms; thus, although the actual length fluctuates only very slightly,[6] the distance between the average positions of the atoms is reduced. For a simple rotation, the ratio is expected to vary as $<\cos\theta>$, where θ is the rotation angle; this provides a good approximation to the observed results. An unconstrained refinement of X-ray data for rubredoxin[7,8] did in fact lead to a wide range of bond lengths. Although it is probable that this study does not present accurate results for the effective bond lengths because the data were too limited for such an unconstrained refinement, it is nevertheless of interest.

Most protein crystallographic refinements (even with relatively high-resolution data) use a model with many assumed parameters,[9] so that the results obtained are going to be biased by that model. A related problem concerns the nonbonded interactions, which are required to be in the attractive region in many refinement programs. Here again, the average structure can have nonbonded contacts that would be strongly repulsive if they occurred in any of the individual structures over which the average is taken.

Although the dynamic simulations that have been used may be significantly in error because of the approximations in the potential function and the neglect of solvent and the crystal environment, they make it clear that there are likely to be problems with some of the crystallographic data that are being treated as the "ultimate truth" in studies of macromolecules.

MOTIONS IN PROTEINS

Rather than giving a general introduction to molecular dynamics of macromolecules (see ref. 10), I shall focus on some specific aspects based on recent developments in the field. One of these is concerned with the time scales of the motions contributing to the observed root-mean-square atomic fluctuations and the other with a harmonic analysis of protein dynamics.

The motions that occur in proteins cover a vast range of time scales (see TABLE 3). Some occur in the femtoseconds range (*e.g.*, those involving bond stretching) whereas others go over periods of seconds or longer (*e.g.*, ring flipping of aromatic residues), so that the method that is to be used for

TABLE 3. Classification of Internal Motions of Globular Proteins

Scales of motions (300°K)	
Amplitude	0.01–100 Å
Energy	0.1–100 kcal
Time	10^{-15}–10^3 s

	Types of motions
Local	atom fluctuations, side chain oscillations, loop and "arm" displacements
Rigid body	helices, domains, subunits
Large-scale	opening fluctuation, folding and unfolding
Collective	elastic-body modes, coupled atom fluctuations, soliton and other nonlinear motional contributions

studying the dynamics has to be adapted to the problem. The folding and unfolding of proteins is a dynamical phenomenon that is most difficult to treat theoretically because it is both complicated and slow. Other problems such as enzyme kinetics, which are also slow macroscopically, involve individual events (*i.e.*, barrier-crossing trajectories) that are fast so that they can be studied by specialized molecular dynamics simulation methods.[11] For "brute force" simulations, the record at the moment is held by a 300 ps simulation of myoglobin that was done on a VAX 11-780 by Dr. R. Levy at Rutgers.[12] With supercomputers an increase of one or two orders of magnitude is possible but still longer simulations will have to wait for new, perhaps specialized computers.

Although considerable experimental information from X-ray crystallography is available on the magnitude of atomic fluctuations in proteins,[13] much less is known about their time scale. Molecular dynamics simulations have been employed in an analysis of the time dependence of the fluctuations.[14] A 25 ps trajectory[15] for BPTI was used to obtain subaveraged root-mean-square (rms) fluctuations; that is, the entire trajectory was divided into a series of intervals of a given length, the mean-square displacements relative to the mean for each interval determined and the results averaged for the entire trajectory. FIGURE 2 shows the results for the C^α atoms. There is a significant contribution to the rms fluctuations from the subpicosecond motion. At 0.2 ps, the C^α rms averaged over all residues is already 0.13 Å, about 40% of the value at 5.0 ps. The average rms value increases slowly with time during the trajectory. Even at 25 ps, the length of the simulation, the asymptotic value for the average rms fluctuation may not have been reached for all atoms, though the results for most atoms have converged.

Comparing the relative values of the C^α fluctuations, we see that at 0.2 ps the results are rather uniform. Since local high-frequency oscillations make the main contribution to this subaverage, it appears that the effective potential does not vary significantly throughout the protein. This is in accord with expectations, if the dominant factor for the high-frequency oscillations is the librational potential associated with torsional motion of the backbone atoms. It should be noted, however, that even on the 0.2 ps time-scale there

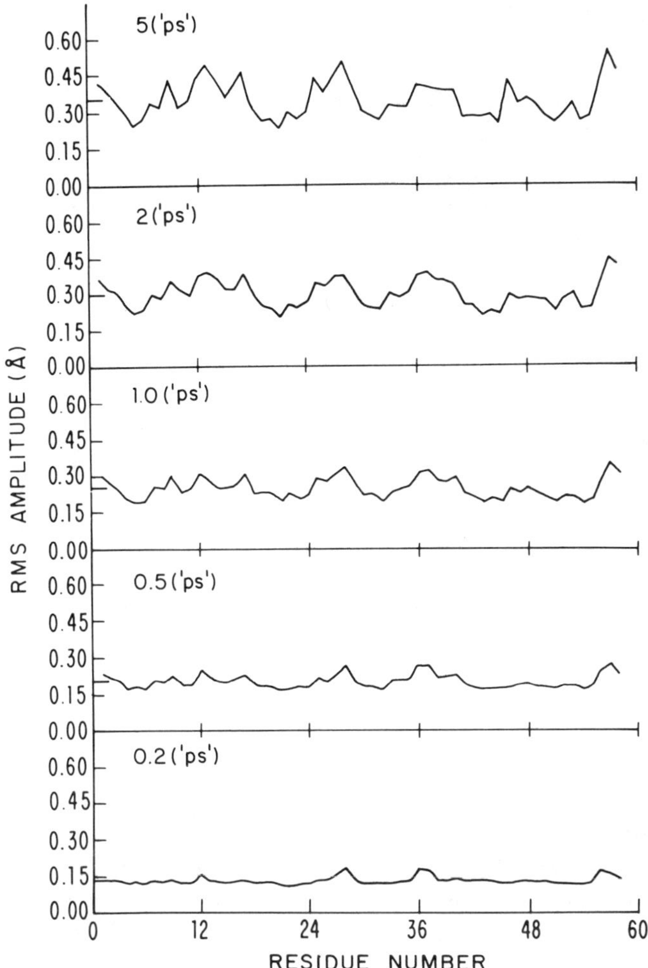

FIGURE 2. Root-mean-square displacement subaverages as a function of residue number for the C^α atoms (T_{av} = 0.2, 0.5, 1, 2 and 5 ps).

is some suggestion of inhomogeneity, in that in the neighborhood of atoms 12, 28, and 36, as well as of the *N*- and *C*-terminal ends, slightly larger fluctuations are found; these regions all have greater than average fluctuations when the low-frequency modes are included (as can be seen from a comparison of the 0.2 and 5.0 ps values). From the time series results (FIG. 3), the 0.5 ps averages include all the high-frequency contributions and the lower frequency motions are becoming more important. This is clearly indicated in FIGURE 2 by the greater variation in the fluctuations along the polypeptide chain. As longer time subaverages are examined, we see that the magni-

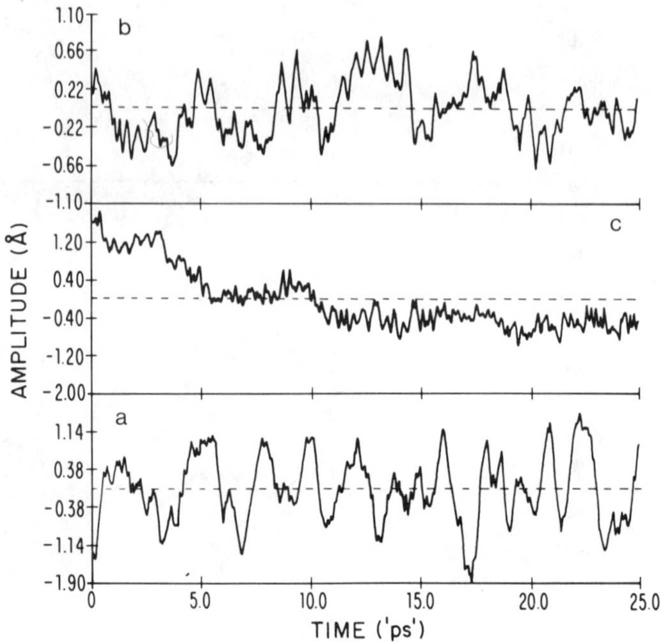

FIGURE 3. Time series for largest Cartesian component in principal axis system of thermal ellipsoid: **(a)** Tyr 21 $C^{\delta 2}$, **(b)** Asp 50 C^{β}, and **(c)** Lys 15 C^{δ}.

tudes of the fluctuation continue to increase in certain regions, in accord with the longer relaxation times and lower frequency, collective character of the modes involved; examples (see FIGURE 4) are the loop region (residues 24–28) and the region at the top of the molecule (around residues 14 and 38).[16] By contrast, for other parts of the protein (*e.g.*, the β-sheet region, 18–24), the C^{α} fluctuations have already reached their asymptotic value by 2 ps.

Time series and time development of the mean-square displacements, as well as correlation functions for the atoms of BPTI,[14] suggest that the fluctuations generally involve the superposition of two types of motions. One is a high-frequency oscillation of relatively small magnitude, and the other has a considerably lower frequency and larger amplitude. From the characteristics of the individual atom fluctuations, and from the relation between the displacements of different atoms, we can draw qualitative conclusions concerning the nature of the two types of motional contributions. The high-frequency oscillations are local, in the sense that they correspond to the librational motion of the individual atoms in an effective potential. By contrast, the lower frequency components have a nonlocal, more collective origin, in that they involve the correlated motion of groups of atoms, ranging from a small number of atoms next to each other along the backbone and/or along a single side chain to a much larger number in certain regions of the protein.

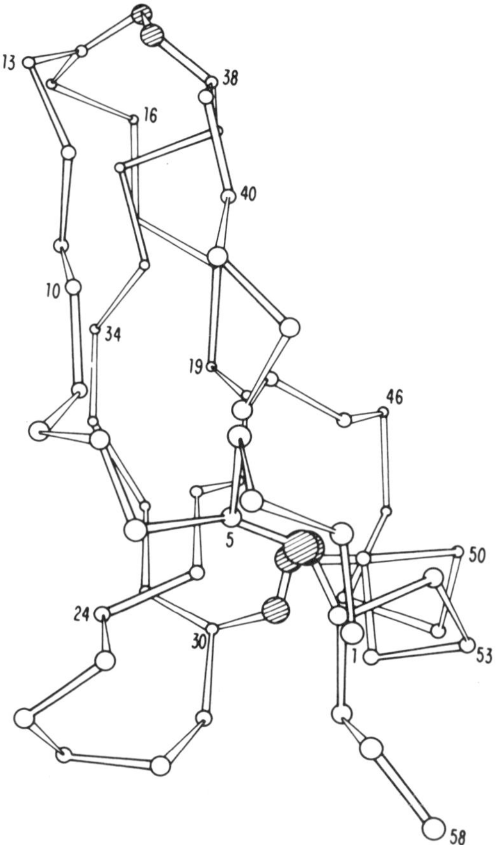

FIGURE 4. X-ray structure of BPTI 16; only α-carbons and disulphide bridges are shown.

The frequencies for the collective mode contributions to the correlation functions vary from $\sim 1 \times 10^{11}\text{s}^{-1}$ to $\sim 1 \times 10^{12}\text{s}^{-1}$; in wave numbers this corresponds to values between 3 and 30 cm^{-1}. This range is in accord with the lowest normal mode frequencies obtained for BPTI (see below).

The results of this analysis, which suggest that protein atom fluctuations have two types of components, are of general interest. The collective, lower frequency modes may well be of particular importance in the biological function of the protein; *e.g.,* they may be involved in the displacements of side chains, loops or other structural units that are required for the transition from an inactive to active configuration of a globular protein and in the correlated fluctuations that may play a direct role in enzyme catalysis. Further, the extended nature of these motions makes them more sensitive to the environment (*e.g.,* the observed differences in simulation results between a vacuum and a solvent run). Because they involve sizeable portions of the surface of the protein, the collective motions are expected to be quenched at low temperature by freezing of the solvent. Their contribution to the mean-square fluctuations may explain the transition observed near 200°K in the temperature dependence of fluctuations in proteins such as myoglobin.[17]

HARMONIC MODEL FOR MOTIONS

An alternative approach to the dynamics of a protein or one of its constituent elements (*e.g.*, an α-helix) is to assume that the harmonic approximation is valid. Early attempts to examine dynamical properties of proteins or their fragments used the harmonic approximation. They were motivated by vibrational spectroscopic studies,[18] where the calculation of normal mode frequencies from empirical potential functions has long been a standard step in the assignment of infrared spectra.[19] In calculating the normal vibrational modes of a molecule, one assumes that the vibrational displacements of the atoms from their equilibrium positions are small enough that the potential energy can be approximated as a sum of terms which are quadratic in the displacements. The coefficients of these quadratic terms form a matrix of force constants which, together with the atomic masses, can be used to set up a matrix equation for the vibrational modes of the molecule.[19] Solution of this equation generally requires the diagonalization of a $3n$-dimensional matrix, where n is the number of atoms in the molecule. The result is obtained as a set of $3n$ eigenvalues (vibrational frequencies) and $3n$ eigenvectors (normal modes). Six of these eigenvectors are associated with eigenvalues of zero, corresponding to overall translation and rotation of the molecule. The remaining $3n - 6$ eigenvalues are the internal vibrational frequencies of the molecule; the associated eigenvectors give the directions and relative amplitudes of the atomic displacements in each normal mode.

Although the harmonic model does not provide a complete description for the motional properties of a protein because of the contribution of anharmonic terms to the potential energy, it is nevertheless of considerable importance because it does serve as a first approximation for which the theory is highly developed. Further, the harmonic model is essential for quantum mechanical treatments of vibrational contributions to the heat capacity and free energy.[19,20]

A harmonic dynamic analysis of the BPTI using the CHARMM program potential function has recently been performed in the full conformational space of the molecule[22]; that is, all bond lengths and angles, as well as dihedral angles, are included for the 580 atom system consisting of all heavy atoms and polar hydrogens.

The histogram in FIGURE 5 shows the calculated normal spectrum: (a) lists all the frequencies up to 2000 cm⁻¹ (hydrogen stretching frequencies are not shown), (b) gives the cumulative distribution for the number of modes below a given frequency, and (c) shows an expanded cumulative distribution for the lowest 300 modes of primary interest. There is an essentially continuous, though not completely uniform, distribution of frequencies between

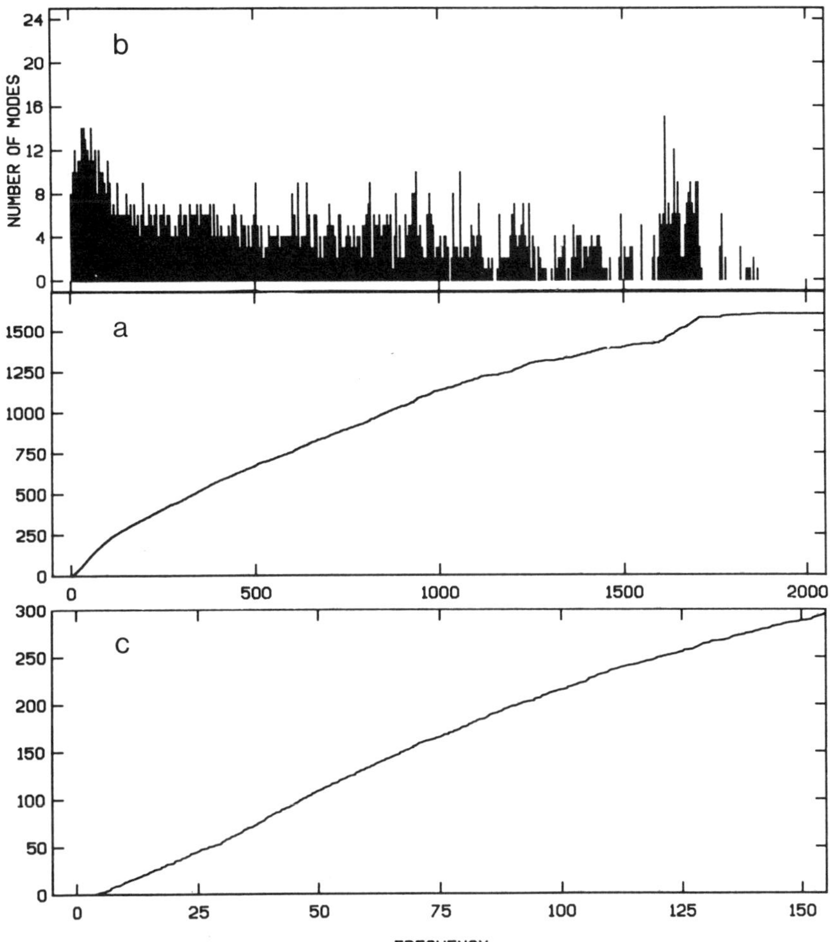

FIGURE 5. Normal frequencies: **(a)** histogram of the number of normal modes per 5 cm⁻¹ interval (hydrogen stretches are not shown); **(b)** number of modes below a given frequency; **(c)** expanded version of **(b)** in the low-frequency region.

3.1–1200 cm⁻¹. Between 1200–1800 cm⁻¹, the frequencies tend to come in groups, many of which are dominated by bond-stretching vibrations. There are twenty modes between 3.1–13 cm⁻¹ and there is a peak in the frequency distribution near 50 cm⁻¹. Because the structure used was not an absolute minimum, seven negative modes were found; energy searches along these modes, which are all local in character, indicated that their correct frequencies are in the range 20–40 cm⁻¹.

The root-mean-square (rms) atom fluctuations were calculated from the normal modes by evaluation of the classical expression[20]

$$\langle \Delta r_k^2 \rangle = k_B T \sum_i \frac{|\vec{a}_{ik}|^2}{\omega_i^2}$$

where $\vec{a}_{ik}$ is the vector of the projections of the ith normal mode with frequency ω_i on the Cartesian components of the displacement vector for the kth atom, k_B is the Boltzmann constant and T is the absolute temperature; quantum corrections are negligible above $50°K$.[20] FIGURE 6 shows the normal mode rms fluctuations calculated at $300°K$ and compares them with the results of a molecular dynamics simulation of PTI in a van der Waals solvent[15]; this simulation was used because its average structure is closest to that employed for the normal mode analysis. The results for main chain and side chain averages as a function of residue number are given (see also TABLE 4). For the main chain fluctuations, the molecular dynamics and normal mode values are very similar; for the side chains there is also a correspondence, though the differences are more pronounced. The main chain values show that the C-terminal end has large fluctuations, as does the loop region at the bottom of the molecule (residues 25–29) and the binding site in the neighborhood of residues 14 and 38 at the top of the molecule. By contrast the β-sheet residues (18–24, 29–35) show smaller fluctuations; the α-helices (3–7, 47–56) are intermediate (see FIGURE 4).

The origin of the differences between the molecular dynamics and normal mode results is likely to have contributions from anharmonic and solvent effects and from the difference between the average dynamics structure and that used for the normal mode analysis. The main chain atoms apparently experience a potential of mean force that is closer to the harmonic potential than do the side chains; since the dynamics simulation was done in a van der Waals solvent, the exterior side chains are expected to be most perturbed. To investigate the nature of the potential in the neighborhood of the structure used for the harmonic model, we have done an energy search along the low-frequency normal-mode displacements. The energy dependence showed significant anharmonic contributions. Fitting the resulting energies to a parabola generally led to a small increase in the effective frequencies ("adjusted" frequencies) which reduced the rms fluctuations; the relative values of the fluctuations are essentially unchanged.

TABLE 4. Root-Mean-Square Fluctuations (Å)

Type	Normal modes	Dynamics[a]
All atoms	0.776	0.714
Main chain	0.598	0.582
Side chain	0.905	0.846
C^α	0.611	0.536
C^β	0.712	0.662
γ position	0.736	0.768
δ position	0.841	0.912
ε position	0.996	0.970

[a] From ref. 22.

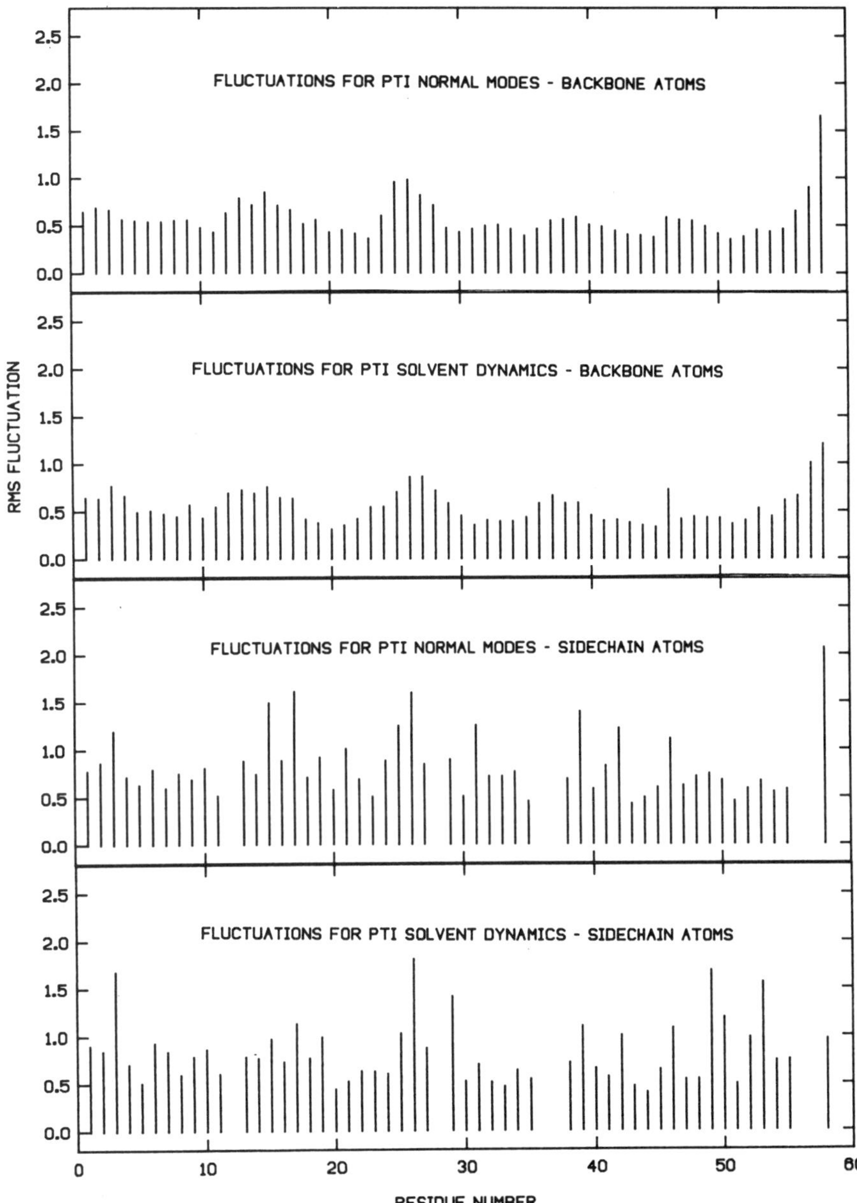

FIGURE 6. Root-mean-square atomic fluctuations (Å) at 300°K averaged over each residue from normal mode and molecular dynamics; separate plots are given for main chain (N, C^α, C) and side chain heavy atoms.

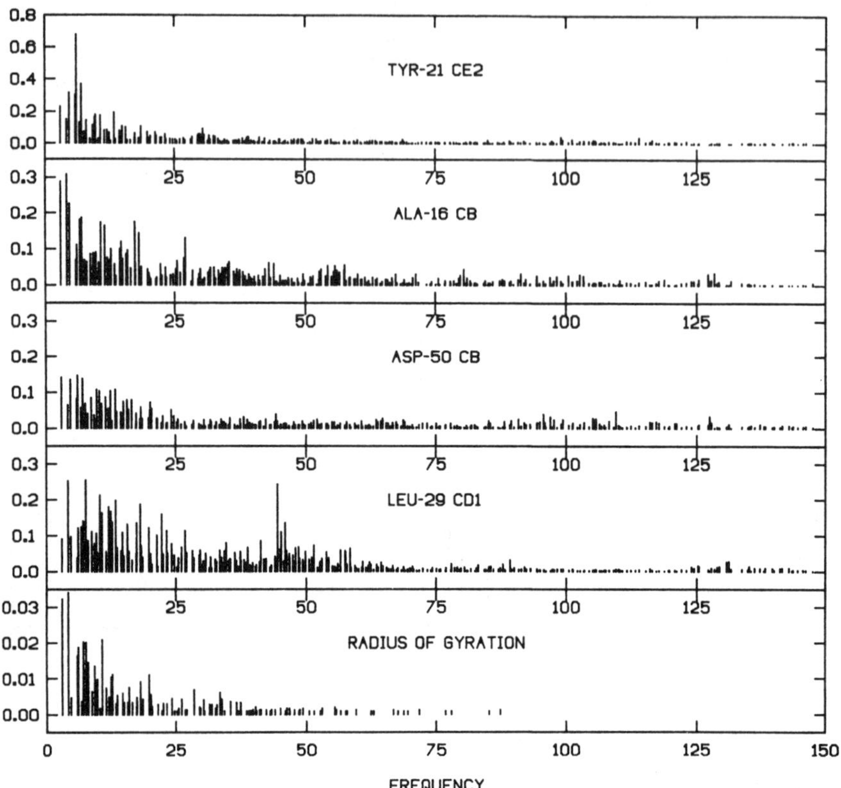

FIGURE 7. Contribution of normal modes to rms fluctuations (Å) as a function of frequency; selected atoms and the radius of gyration are included.

Analysis of the time scale as well as the magnitude of the fluctuations has been made for the molecular dynamics results. From the calculated time series and correlation functions (see above), it has been found that the atomic motions contributing to the rms displacements generally have a small local high-frequency component ($\sim$0.2 ps or $\sim$150 cm^{-1}) on which are superposed motions of a more collective character with time scales ranging from 1–10 ps or 30–3 cm^{-1}. The present normal mode study essentially confirms the dynamics results. FIGURE 7 shows the contributions of the different normal modes to the displacements of some of the atoms whose motions were analyzed in the molecular dynamics simulations; also included is the fluctuation of the radius of gyration for the molecule. In most cases, the dominant contributions come from low-frequency modes in the range 3–50 cm^{-1}, although non-negligible contributions come from higher frequencies up to 130 cm^{-1}. It is evident that for certain atoms (*e.g.,* Tyr 21 C$^{\epsilon 2}$), only a very small number of modes are important, whereas for other atoms (*e.g.,* Ala 16 C$^\beta$, Asp 50 C$^\beta$)

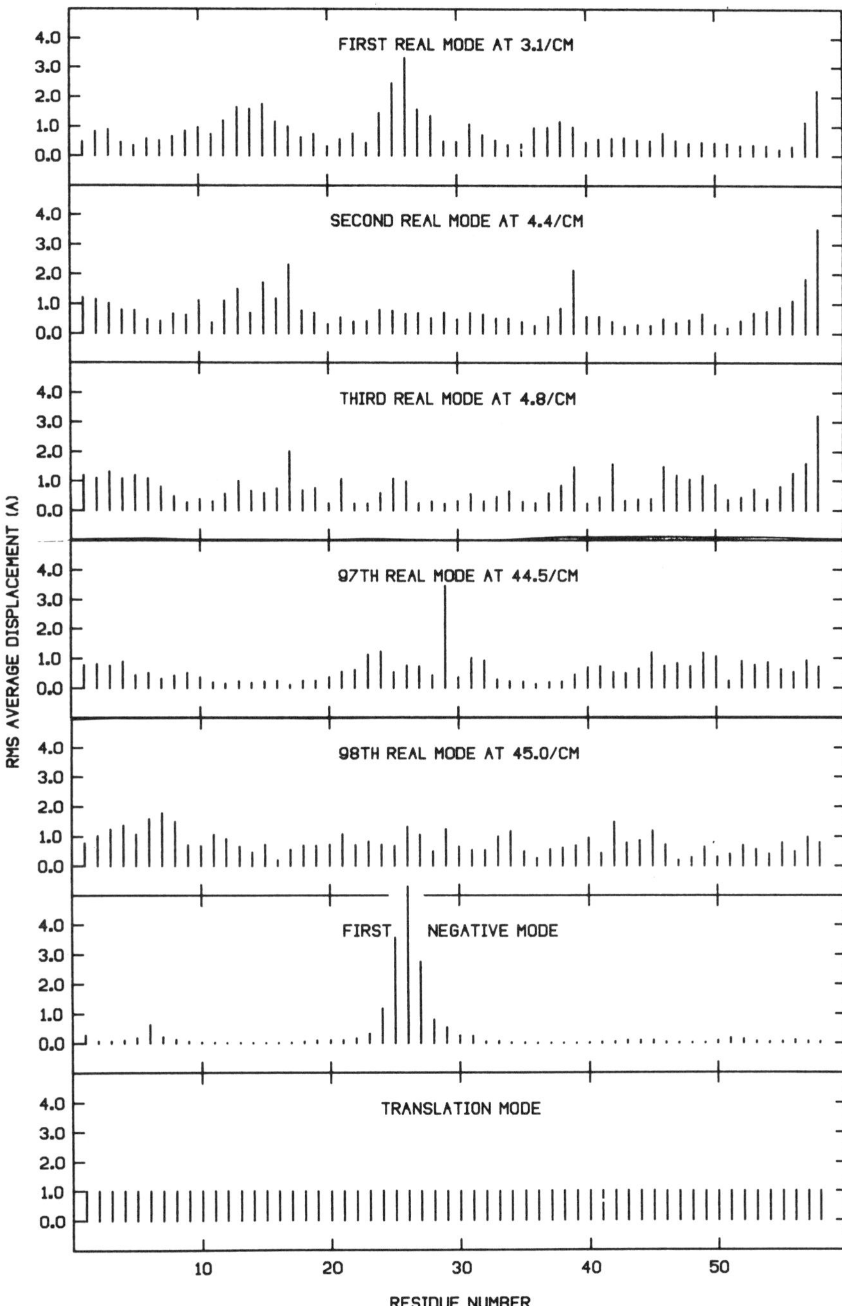

FIGURE 8. Normal mode distribution: rms average displacement of atoms within a residue for a 1 Å rms displacement along selected modes.

a range of frequencies are involved; for Leu 29 $C^{\delta 1}$, a mode at 44.5 cm^{-1} makes a large contribution.

It is of considerable interest to examine the form of the normal modes themselves. This is of particular importance for the evaluation of the correlation between the motions of different atoms and different groups of atoms. Analysis of the dynamics results has indicated that the larger scale motions have a collective character that may involve a few neighboring atoms, a residue, or groups of many atoms in a given region of a protein. We show in Figure 8 the distribution of the displacements over residues of some of the low-frequency modes. Also included is one of the translation modes, which clearly demonstrates the purity of this mode, a pictoral verification of the accuracy of the normal mode determination procedure. Most of the 300 lowest modes are highly delocalized; they are generally distributed over the entire molecule. A striking exception is one of the "negative" modes that is localized in the loop region. The lowest real mode (at 3.1 cm^{-1}) mirrors the overall rms fluctuations (see Figure 6). Other modes shown, although they are also delocalized, are distributed somewhat differently over the various portions of the molecule. In considering the character of the individual modes, it must be recognized that because of the close spacing, relatively small effects, such as solvent damping or external perturbations (*e.g.,* ligand binding), can lead to significant mode mixing. This may be of biological interest. It also suggests that rather than individual mode properties, those that involve averages over a range of modes with similar frequencies are likely to be most significant and least sensitive to anharmonic corrections.

From Figures 7 and 8 it is clear that different residues make varying contributions to the different modes, and *vice versa*. This leads to the speculation that certain mutations may effect the motions of proteins in specific ways. Obviously structural changes may also be involved, but it is suggestive of what it may be possible to do from a dynamic viewpoint by site-specific mutagenesis.

Finally, let me turn to a point concerned with the question of ligand binding that has been made clearer by the normal mode analysis. It has been suggested that binding of ligands and substrates can have a significant effect on thermodynamic properties by perturbing the low-frequency vibrations of the protein; *e.g.,* for BPTI some change might be expected on binding to trypsin. As a model for this effect, we compare the results obtained for the directly calculated and "adjusted" modes (see above), the latter corresponding to the "perturbed" system with somewhat higher frequencies. At 100°K, the vibrational free energy changes from -41.5 to -37.7 kcal/mol in the presence of the perturbation; at 300°K, the values are -336.4 and -325.1 kcal/mol, respectively. At all temperatures the vibrational enthalpy increases and there is a destabilizing effect on the system. The change in enthalpy contrasts with that assumed in previous discussions,[23] due to the fact that we have included the zero-point contribution.

REFERENCES

1. PETTITT, B. M. & M. KARPLUS. J. Am. Chem. Soc. In press.
2. BROOKS, B. R., R. E. BRUCCOLERI, B. D. OLAFSON, D. J. STATES, S. SWAMINATHAN & M. KARPLUS. 1983. J. Comp. Chem. **4**: 187.
3. YU, H. & M. KARPLUS. Submitted.
4. VAN GUNSTEREN, W. F. & M. KARPLUS. 1982. Macromolecules **15**: 1528.
5. BUSING, W. R. & H. A. LEVY. Acta Crystallogr. 1964. **17**: 142.
6. MCCAMMON, J. A., B. R. GELIN & M. KARPLUS. 1977. Nature **267**: 585.
7. WANTENPOUGH, K. D., L. C. SICKER & L. H. JENSEN. 1979. J. Mol. Biol. **131**: 509.
8. WANTENPOUGH, K. D., L. C. SICKER & L. H. JENSEN. 1980. J. Mol. Biol. **138**: 615.
9. KONNERT, J. H. & W. A. HENDRICKSON. Acta Crystallogr. 1980. **A36**: 344.
10. KARPLUS, M. 1981. Ann. N. Y. Acad. Sci. **367**: 407–418.
11. NORTHRUP, S. H., M. R. PEAR, C-Y. LEE, J. A. MCCAMMON & M. KARPLUS. 1982. Proc. Nat. Acad. Sci. **79**: 4035.
12. LEVY, R. M., R. P. SHERIDAN, J. W. KEEPERS, G. S. DUBEY, S. SWAMINATHAN & M. KARPLUS. Biophys. J. To be submitted.
13. PETSKO, G. A. & D. RINGE. 1984. Ann. Rev. Biophys. **13**: 331.
14. SWAMINATHAN, S., T. ICHIYE, W. F. VAN GUNSTEREN & M. KARPLUS. 1982. Biochemistry **21**: 5230.
15. VAN GUNSTEREN, W. F. & M. KARPLUS. 1982. Biochemistry **21**: 2259.
16. DEISENHOFER, J. & W. STEIGEMANN. 1975. Acta Crystallogr. **B31**: 238.
17. PARAK, F., E. N. FROLOW, R. L. MOSSBAUER & V. I. GOLDANSKII. 1981. J. Mol. Biol. **145**: 825.
18. MIYAZAWA, T. 1967. *In* Poly-α-Amino Acids. G. D. Fasman, Ed. Marcel Dekker. New York.
19. WILSON, E. B., J. C. DECIUS & P. C. CROSS. 1955. Molecular Vibrations. McGraw-Hill. New York.
20. LEVY, R. M., D. PERAHIA & M. KARPLUS. 1982. Proc. Nat. Acad. Sci. U.S.A. **79**: 1346.
21. GO, N. & H. SCHERAGA. 1976. Macromolecules **9**: 535.
22. BROOKS, B. & M. KARPLUS. 1983. Proc. Nat. Acad. Sci. **80**: 6571.
23. STURTEVANT, J. M. 1977. Proc. Nat. Acad. Sci. U.S.A. **74**: 2236.

Modeling the Mechanism of Peptide Cleavage by Thermolysin

DAVID G. HANGAUER, PETER GUND,
JOSEPH D. ANDOSE, BRUCE L. BUSH,
EUGENE M. FLUDER, EUGENE F. McINTYRE,
AND GRAHAM M. SMITH

Merck Sharp & Dohme Research Laboratories
Rahway, New Jersey 07065

THE MERCK MACROMOLECULAR MODELING GRAPHICS FACILITY[a]

Introduction

Merck & Co. has for some time utilized a molecular modeling system that we developed for aiding drug design.[1,2] More recently, it became clear that more powerful systems were needed to study the relevant larger structures which were becoming increasingly available, and to study steric and electrostatic surfaces. We have obtained such equipment and utilized it for drug design. At the same time, realizing that many of the modeling functions can be performed on less expensive equipment, we have modified our program to run on several types of graphics terminals. Characteristics of our current system for molecular modeling, and some applications to drug design, are discussed below.

Merck Molecular Modeling System Capabilities

The basic hardware is shown in FIGURE 1. Whereas our original modeling software ran on IBM equipment under TSO,[1] our current MOLEDIT (MOLecule EDITor) program resides on our VAX 11/780 computer in order to provide more responsive service and a "friendlier" environment for our chemist users. The IBM 3081 is still accessed for cpu-intensive jobs; background jobs are submitted to the IBM computer automatically from the VAX using HASP-plus. The VAX computer is connected by high-speed data link to our Evans & Sutherland Multi-Picture System (MPS) to ensure rapid response, whereas other terminals communicate to the VAX or other computers over dial-up ports.

---

[a] This section was presented by Dr. Gund.

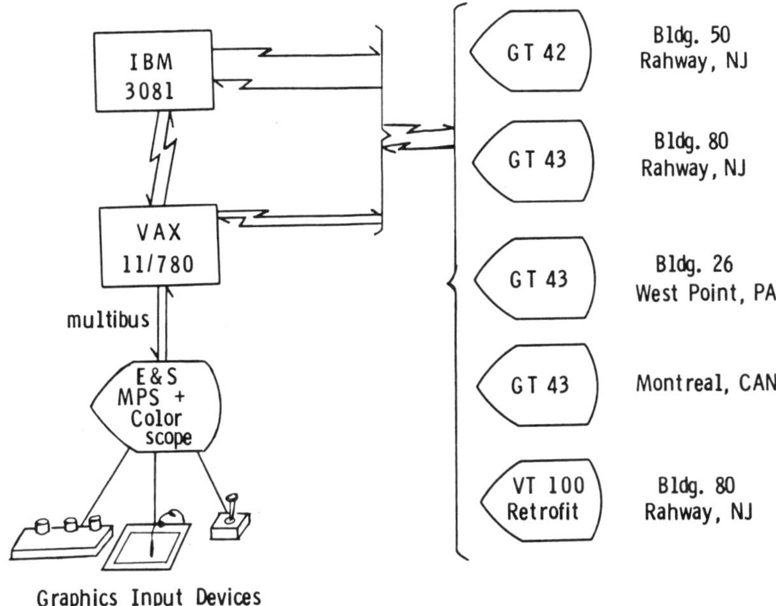

FIGURE 1. Merck molecular modeling system network.

In the past, as technology changed we converted our program to display molecules first on Tektronix terminals (PLOT-10 software), then on DEC GT40-type terminals (GT40LIB software), and finally on the MPS (using E&S-supplied graphics routines or GRAMPS). Graphics terminals are evolving at a rapid rate, and we want to be able to take advantage of new capabilities with minimal software redevelopment. Furthermore, scientists who have obtained specific types of terminals for other applications would like to use those terminals for molecular modeling. Consequently we have reimplemented MOLEDIT utilizing a device-independent graphics package, DI-3000. In this approach, programs are written by calling device-independent graphics routines; various device drivers are then invoked to access the appropriate graphics terminals. Currently we have users running molecular modeling studies on VT100/102 retrographics and VT125 terminals, and on DEC PC350 personal computers.

Our initial modeling system consisted of a series of programs that communicated by reading and writing compatible files of molecular coordinates. In our current system, MOLEDIT acts as a high-level executive program which invokes various subprocesses; when the subprocess is completed, control returns to MOLEDIT. Current capabilities, arranged by modeling function in FIGURE 2, will now be briefly described.

For medium-sized molecules (not enzymes), structures are created by inputting crystal coordinates (XTAL); by inputting standard bond lengths, angles and dihedral angles (COORD); by drawing the structure on the screen

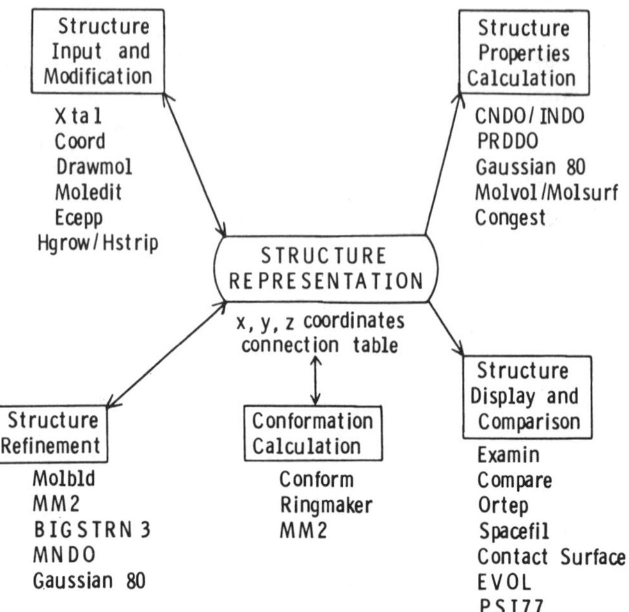

FIGURE 2. Merck molecular modeling system components.

(MOLEDIT:DRAW) and performing an approximate geometry optimization; by inputting polypeptide residue codes and torsion angles (ECEPP); or by modifying or merging previously formed molecules (MOLEDIT edits molecular files much as a text editor modifies or merges text files). More accurate geometries may be derived by a variety of classical mechanical and quantum mechanical programs. Conformations may be generated systematically by several procedures, including Professor Clark Still's RINGMAKER program[3] and an improved version developed at Merck. Computer graphics programs are used to display molecules as stick, ball and stick, and spacefilling figures, as contact surface representations, or as superpositions. Our implementation of excluded volume maps[4] and orbital and electrostatic field contours may also be represented. Physicochemical properties, such as frontier orbitals and charge densities, molecular volume and surface area, may be calculated from the molecular coordinates.

Macromolecular Modeling System

The same VAX and MPS hardware, but different software, are used for macromolecular modeling[5] (FIGURE 3). Special-purpose programs written by one of us (GMS) may be used to carve out active sites or sections of enzyme structures, which may then be displayed using MOLEDIT, FRODO, or GRAMPS. Alternatively Dr. T. Alwyn Jones's FRODO program, extensively

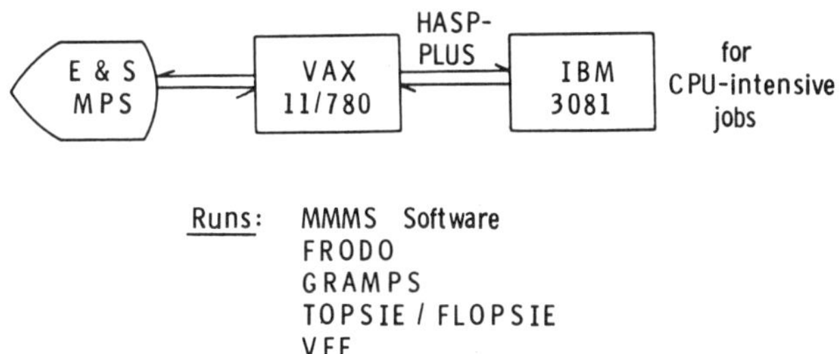

FIGURE 3. Merck molecular modeling graphics facility (MMGF).

modified by one of us (BLB), may be used interactively to study an enzyme and write a file to be utilized by MOLEDIT. Useful features added to FRODO include a novel method of dynamically highlighting intermolecular contacts during the "steering" of a substrate or inhibitor into an enzyme active site.[5]

Modeling the interaction of small (drug-sized) and large (enzyme-sized) molecules requires software that spans two separate disciplines. The small molecule modeling arose from the traditions of physical organic chemistry, fed by theoretical chemistry; it typically is concerned with very careful calculations to "explain" chemical phenomena hanging on differences of a very few kilocalories. Macromolecular modeling, on the other hand, derives from the discipline of biophysics and especially the subdiscipline of protein crystallography; the very large molecules of interest typically are composed of common subunits or "residues" (such as the 20 naturally occurring amino acids), and energies are computed by more approximate methods involving movement of groups of atoms. In order to address problems at the boundary of these two domains, we have added knowledge of atom-centered information to macromolecular modeling programs such as FRODO, and added knowledge of residue information to MOLEDIT. The optimal software for these applications is not yet in hand, however.

Whereas graphical examination of enzyme-inhibitor interaction may provide qualitative indication of binding strength, semiquantitative or quantitative estimation requires some sort of energy calculation. Given the number of atoms involved in an enzyme-inhibitor complex, it is useful to be able to make some simplifying approximations and assumptions. An approach to this problem has been made by Professor Tom Halgren of City University of New York during a sabbatical year at Merck, who has written an experimental program called TOPSIE (Torsional Optimization Package for Substrates and Inhibitors of Enzymes) that optimizes the geometry of substrates in their bound conformation by varying torsion angles. Preliminary results with TOPSIE were encouraging, but suggested that fully optimized structures are needed for valid comparisons of energies. An enhanced version of the program, OPTIMOL, allows such full geometric optimization of the inhibitor in the active site, and gives better energies. For selected cases, full geometry optimization of the substrate plus active site side chains, or ultimately of substrate plus entire enzyme, would be desirable; FLOPSIE is being extended in these directions.

Pro-Phe-D-Trp   H-D-Phe-Cys-Phe-D-Trp

Phe-Thr-Lys   Thr(ol)-Cys-Thr-Lys

(1)   (2)

(3)

Molecular/Macromolecular Modeling System Applications

Molecular and macromolecular modeling studies may be conducted to support current products; to optimize lead compounds; and to generate new leads. Current products are supported by mechanism of action studies; by comparison to related drug structures; by determination of the bioactive conformation; by aiding spectral interpretation; and by computing reactivity and reaction stereoselectivity. Analog optimization projects may be aided by mechanism or mode of action modeling studies; by conformational analyses; by the use of conformation or shape parameters for QSAR studies; by generating and testing structure-related hypotheses; by deriving pharmocophoric patterns; by modeling receptor active sites; and by comparing electrostatic field surfaces. In order to generate new lead compounds, modeling studies may focus on mechanism of enzyme action; mechanism of action of bioactive metabolites and competitive drugs; modeling new species containing a desired pattern; modeling minimal essential functionality of a hypothetical receptor or binding site; modeling the interaction of model enzymes with hypothetical inhibitors; building a model of a target enzyme structure by transforming the sequence of a related, solved enzyme crystal structure; modeling other active sites such as receptors, ion channels, and DNA intercalation sites; aiding the structural elucidation of microbiological leads; and modeling reactant-like enzyme inhibitors, transition state inhibitors, and collected product inhibitors.

Several examples of molecular modeling applications to aid drug design are given in the earlier references.[1,2] More recently, conformational studies aided the development of a potent cyclic hexapeptide analog (1) of the metabolically labile tetradecapeptide somatostatin.[6] Modeling studies at Sandoz in Basel were said to have aided development of a different octapeptide analog of somatostatin (2), which is also being studied in the clinic.[7] In another example, conformational studies on substrate intermediates and inhibitors aided the development of nonpeptidic angiotensin converting enzyme inhibitors, *e.g.,* (3), as antihypertensives.[8]

Macromolecular modeling has already been effectively utilized in a number of projects. For example we have used it to study the dihydrofolate crystal structures of Matthews and co-workers,[9] although far more work along these lines has been performed at Burroughs-Wellcome[10] and at the University of California at San Francisco Graphics Laboratory. The systems have been usefully applied to the problem of designing inhibitors of renin.[11] More recently the system has been used for studying the hydrolytic mechanism of thermolysin, as described below, and is presently being utilized to design inhibitors of this enzyme.

MODELING THE MECHANISM
OF PEPTIDE CLEAVAGE BY THERMOLYSIN[b]

For the utility of macromolecular modeling to become fully appreciated, and to guide its development further, this new tool must be applied to important scientific problems. The availability of high-resolution crystallographic coordinates for a number of native enzymes and inhibitor complexes provides an opportunity to apply macromolecular modeling to the study of enzyme reaction mechanisms. Theoretical studies of this type can provide extremely useful background for understanding crystallographically solved enzyme inhibitor complexes and rationally exploring structural modifications of these inhibitors which might result in enhanced binding. The coordinates generated for the transition state in a modeling study of this type can be used as a "template" after which transition state analog inhibitor designs may be patterned.

In this half of the presentation the application of the Merck Macromolecular Modeling Graphics Facility to studying the mechanism of peptide cleavage catalyzed by thermolysin will be discussed. The reasons for undertaking this study in an industrial setting will be mentioned. The modeling strategy, assumptions and methods used will be detailed and the most significant results highlighted. A more complete account of the results and proposed mechanism, along with color stereo figures of the key modeled complexes, will be given elsewhere.[12]

Reasons for Studying Thermolysin

Thermolysin is a zinc endopeptidase that is particularly suited for a macromolecular modeling study with interactive computer graphics because a variety of crystallographic, inhibitor, substrate and chemical modification data are available. The high-resolution (1.6 Å) crystal structure of the na-

---

[b] This section was presented by Dr. Hangauer.

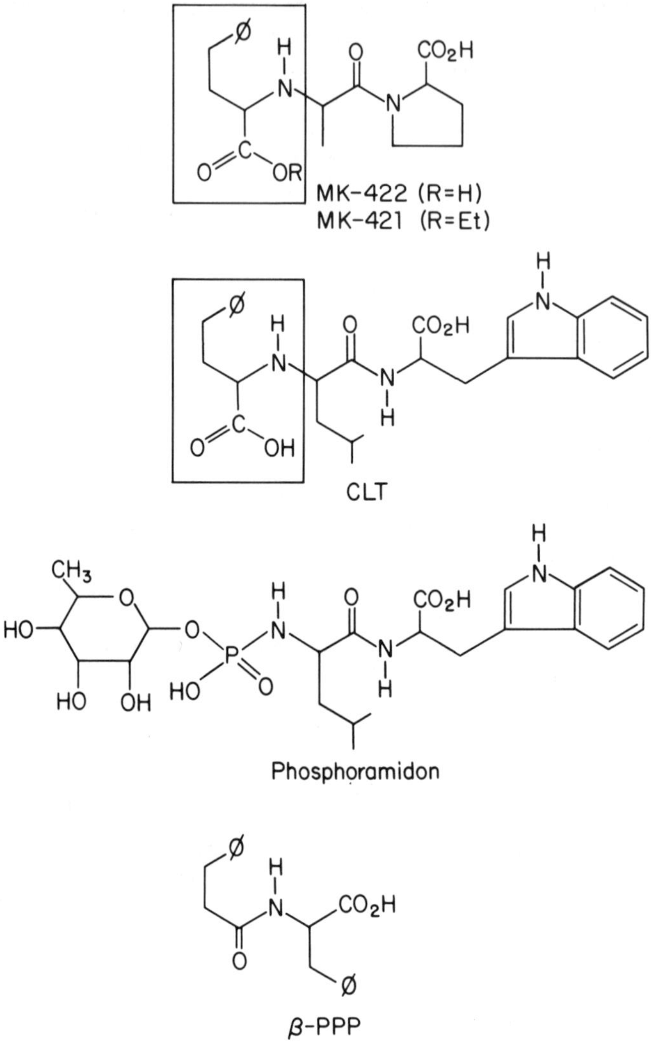

FIGURE 4. Comparison of the thermolysin inhibitor CLT with ACE inhibitor MK-422 and two other thermolysin inhibitors, phosphoramidon and β-PPP.

tive enzyme has been reported.[13] The enzyme is composed of two roughly spherical domains joined by a helix. The active site is in a deep cleft separating the two domains with the zinc atom positioned at the bottom and roughly equidistant from the ends of the cleft. The zinc is held by three ligands from the enzyme (His 146, His 142 and Glu 166) and has one water molecule coordinated to it resulting in a distorted tetrahedral array. The catalytically important residues located near the zinc are His 231 and Tyr 157 on one side of the cleft, and Glu 143 on the other.

Thermolysin is a good model enzyme for mechanistic studies of zinc peptidases in general since a close correspondence exists in the location and function of the mechanistically important residues with various other zinc enzymes.[14,15] The angiotensin converting enzyme (ACE) is an important zinc peptidase involved in the control of blood pressure. This enzyme has not been studied crystallographically, presumably due to difficulties in obtaining suitable crystals. The prodrug MK-421 (FIG. 4) is presently undergoing clinical trials for the treatment of human hypertension. The active drug is MK-422 (FIG. 4) which is a very potent inhibitor of ACE.[16] The successful transfer of the MK-422 inhibitor design from ACE to thermolysin[17] reinforced our interest in thermolysin as a model zinc peptidase. A potent (K_i = 5 × 10^{-8}M) thermolysin inhibitor, N-(1-carboxy-3-phenylpropyl)-L-leucyl-L-tryptophan (CLT), is also shown in FIGURE 4. The X-ray structure of the CLT:thermolysin inhibitor complex has recently been solved[19] to a resolution of 1.9 Å. The boxed fragments in FIGURE 4 indicate the design carryover from the ACE inhibitor MK-422 to the thermolysin inhibitor CLT. The dipeptide region of these inhibitors simply reflects the specificity differences between the enzymes in the S_1' and S_2' subsites.

The mechanism of thermolysin catalyzed peptide cleavage has been studied previously with physical models and unrefined coordinates.[18] However, the X-ray study of the CLT:thermolysin complex yielded the surprising discovery that the zinc binding carboxyl group of CLT coordinates in a bidentate fashion. This result, along with the availability of refined coordinates and the Merck Macromolecular Modeling Graphics Facility, convinced us that a reinvestigation was in order.

Strategy, Assumptions, and Methods

Our modeling study, by necessity, required certain assumptions. The most obvious assumption is that the crystal structure of the enzyme is representative of the solution conformation. We felt reasonably confident in making this assumption because thermolysin is unusually resistant to thermal denaturation[20] and available experimental comparisons[21] of the crystal and solution conformations indicated that no detectable conformational differences exist. Furthermore, peptide hydrolysis has been observed with the crystalline enzyme.[22] Of the numerous thermolysin inhibitor complexes studied crystallographically by Matthews and colleagues, only one[23] showed a significant movement of the enzyme from its native conformation. In this isolated case the movement was attributed to the irreversible nature of the inhibitor (Glu 143 was alkylated) which, by necessity, caused a 0.2–0.3 Å shift of the peptide chain adjacent to Glu 143 along with some other movements.

At the time our modeling study was carried out, the OPTIMOL program was not available for relaxing our modeled complexes to their nearest local minimum energy conformations. Consequently, we relied upon FRODO and the modifications made by Bush,[5] especially CLARIFI, to guide us in positioning our model substrate in the active site. When examining a candidate complex, the distance for each of the inter- and intramolecular contacts was measured and compared to standard van der Waals contact radii.[24] When nonpolar atoms were near van der Waals contact distance from each other, hydrophobic binding was assumed to occur and contribute to stabilizing the complex. When heteroatoms could be positioned within about 2.7–3.0 Å of each other, and a hydrogen atom was present on one of the heteroatoms, hydrogen bonding was assumed to occur and stabilize the complex. Salt bridges were also assumed to provide a stabilizing force for the complex. Although these methods seemed somewhat crude, we felt that they could provide qualitatively accurate results with a minimum of computer time and effort. Furthermore, there is an uncertainty of about ±0.2 Å in the positioning of the enzyme atoms in the X-ray structure and minor adjustments of the enzyme in response to the substrate seem reasonable. The ease with which these methods can be applied allows the user to explore rapidly a large number of candidate complexes.

A requirement in our strategy for this study was that it should make maximal use of the available experimental data. Even the most advanced current methods for estimating the energetics of ligand binding to an enzyme typically require assumptions and approximations that necessitate caution in interpreting the results. An accurate quantitative prediction of the Gibbs energy change associated with ligand binding is difficult to accomplish, perhaps due mainly to uncertainties in accounting for desolvation and entropy changes. Since refined coordinates were available for two very potent thermolysin inhibitors, CLT and phosphoramidon[25] (FIG. 4), along with unrefined coordinates for β-phenylpropionyl-L-phenylalanine (β-PPP, FIG. 4),[18] the peptide sequence Phe-Leu-Trp was chosen for the P_1-P_1'-P_2' region of our model substrate. The Leu-Trp dipeptide fragments of both CLT and phosphoramidon were shown to occupy the S_1' and S_2' subsites.[19] The phenethyl fragments of both CLT and β-PPP were shown to occupy the S_1 subsite. Consequently, the model substrate sequence chosen facilitates the maximal utilization of these crystal structures.

At the time of this study no crystallographic data were available for inhibitors which might occupy the S_2 subsite. Therefore, a Z-Phe was chosen for the P_2 position based upon the substrate data of Morihara and Tsuzuki[25] shown in TABLE 1. The Z-Phe is the most sterically demanding of three P_2 residues shown and would be expected to have the most limited number of plausible (on steric grounds) complexes. Furthermore, by simply ignoring the coordinates for the side chain phenyl group, or the entire side chain, one

TABLE 1. Selected Thermolysin Substrate Data[a] Illustrating the Effect of Side Chain Variations at Position P_2

Peptide	K_M (mM)	k_{cat} (sec^{-1})	k_{cat}/K_M (sec^{-1}mM^{-1})
Z-Gly-Gly↓Leu-Ala	12.2	362	30
Z-Ala-Gly↓Leu-Ala	8.6	5,208	606
Z-Phe-Gly↓Leu-Ala	0.9	446	491

[a] From ref. 26.

could obtain information about complexes with a P_2 Z-Ala or Z-Gly, respectively. We chose not to extend the model substrate further because the substrate data[25] indicated that additional residues have a much less dramatic effect on the reaction kinetics.

The interactive nature of this study allowed us to utilize some basic concepts of enzyme function as a guide. For example, we assumed that the active site of thermolysin has been designed by nature to utilize the intrinsic binding energy[26] available from the substrate groups on either side of the scissile peptide linkage to lower the transition state energy. It is not required, nor necessarily desirable, for the intrinsic binding energy of the substrate or products to be fully utilized in the ground state. In terms of our modeled complexes, it is not necessary that the Michaelis complex involve a complete penetration of the substrate into the active site. We assumed that the tetrahedral intermediate is closest in geometry to the transition state.[27] Therefore, its complex will involve a full penetration into the active site so that maximal utilization of the intrinsic binding energy can occur. At some point along the reaction pathway, we expected the enzyme to use the binding potential of the substrate to force the scissile peptide linkage into a destabilizing environment. This destabilization may be accomplished by a combination of the following: (1) an electrostatic environment more suited to the transition state, (2) appropriately positioned functionality (a H_2O nucleophile, a general base, *etc.*), (3) a loss of entropy by the reacting groups, and (4) perhaps modest geometric strain on the scissile peptide.

The results of a modeling study of this type are influenced by the strategy and assumptions invoked by the user. This is a direct consequence of the interactive nature of the computer programs. One should, therefore, be careful to point out the strategy, assumptions, and methods used as a basis for this type of modeling study.

The model substrate Z-Phe-Phe-Leu-Trp (Fig. 5) was constructed using MOLEDIT. The coordinates for the Leu-Trp portion of the substate were derived from the corresponding coordinates for CLT. The remaining substrate coordinates were entered so as to conform with standard bond lengths and angles. Dihedral angles were chosen arbitrarily since they are adjusted during the docking process. The substrate coordinates were then transferred to FRODO and displayed along with the enzyme active site and the inhibitors (in the background, different color). The use of the Leu-Trp coordinates

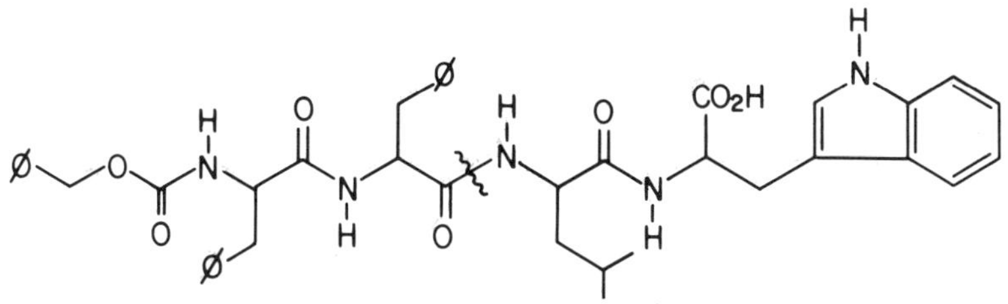

FIGURE 5. Structure of the model substrate Z-Phe-Phe-Leu-Trp used in this study.

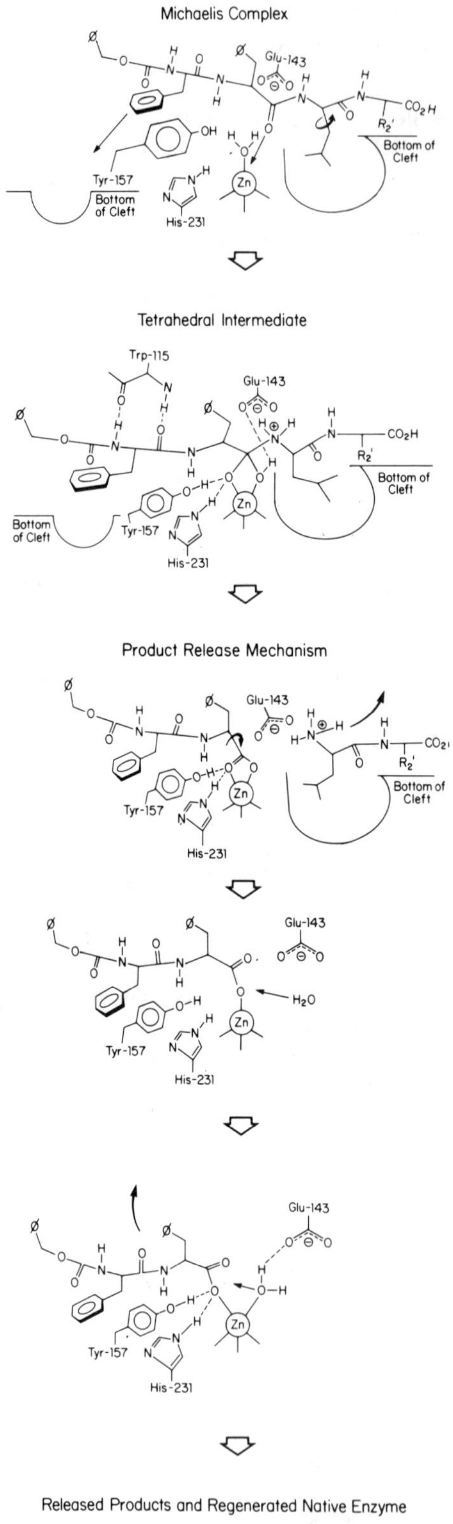

Michaelis Complex

Tetrahedral Intermediate

Product Release Mechanism

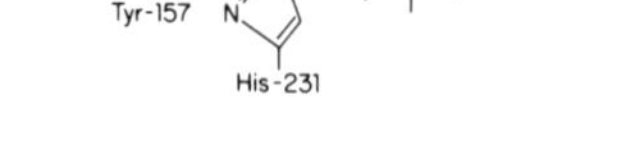

Released Products and Regenerated Native Enzyme

FIGURE 6. Schematic representation of the intermediate complexes and proposed reaction mechanism.

from CLT in constructing the substrate results in an initial docking to the active site.

Results

The first potential Michaelis complex investigated was based upon the thermolysin:β-PPP complex, wherein the β-PPP amide was found coordinating directly to the zinc in place of the water molecule found in the native enzyme. The earlier modeling study[18] has used the thermolysin:β-PPP complex to propose a similar Michaelis complex. We used the coordinates for the zinc chelating amide of β-PPP to guide the positioning of our model substrate scissile peptide group. A superposition of these groups was obtained by making the appropriate dihedral angle changes as well as translations and rotations of the entire substrate. Subsequently, an attempt to obtain a reasonable fit in the active site was made by adjusting the conformation and positioning of the substrate. We found that a serious steric interaction of the S_2 subsite between the P_2 Z-Phe side chain and the bottom of the active site cleft was present. The interaction could not be avoided while maintaining a direct coordination distance between the zinc and scissile peptide oxygen. When we considered this result in light of the basic concepts of enzyme function mentioned earlier, this Michaelis complex seemed less plausible than other possibilities.

We decided to give up the notion that the scissile peptide must coordinate directly to the zinc. Consequently, we backed the scissile peptide, P_1 and P_2 residues out of the active site until the peptide oxygen was 4.2 Å from the zinc. The P_1' and P_2' residues were maintained (at least partially) in the S_1' and S_2' subsites during this motion by appropriate dihedral angle, translational and rotational adjustments of the substrate. A water molecule was inserted as a zinc ligand in the position found in the native enzyme. Final adjustments were made to optimize the fit. The scissile peptide carbonyl was now in a position to hydrogen bond with the zinc-bound water molecule analogous to the outer sphere zinc coordination proposed for an aldehyde carbonyl in liver alcohol dehydrogenase.[28] We found no unreasonable interatomic distances in this Michaelis complex. We noted that the P_2 Z-Phe side chain phenyl group is adjacent and perpendicular to the aromatic ring of Tyr 157 (FIG. 6). Although other Michaelis complex candidates were examined, this is the one we felt was the most likely.

The model substrate Z-Phe-Phe⁻Leu-Trp was converted to the corresponding tetrahedral intermediate via MOLEDIT by adding the elements of water to the scissile peptide (↓) and adjusting the bond angles and lengths to correspond with the standard values for the sp^3 hybridized atoms. The resulting structure was transferred to FRODO and viewed along with the active site and inhibitors. A number of potential tetrahedral intermediate complexes were generated and examined, all of which involved complete penetration into the active site. The earlier modeling study[18] had proposed a tetrahedral intermediate complex based upon the unrefined coordinates for

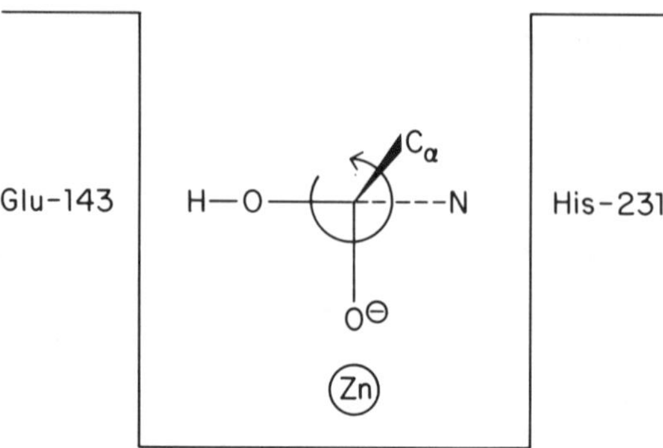

FIGURE 7. Schematic representation of the tetrahedral intermediate orientation analogous to the phosphoramidon:thermolysin complex. The arrow indicates the rotation required to obtain the other orientations investigated.

the thermolysin:phosphoramidon complex. Their modeled complex involved a monodentate coordination of the tetrahedral intermediate to zinc as illustrated in FIGURE 7. We investigated this orientation as well as two other orientations with our model tetrahedral intermediate. The additional orientations are centrally related to each other by a rotation (as shown in FIGURE 7) in the direction of the arrow. In this way, a bidentate zinc coordination and an opposing monodentate coordination was obtained. Of the three possibilities, the bidentate orientation gave the best fit in the active site for the entire substrate. These results, along with the finding that CLT and hydroxamic acid containing inhibitors[22] complex in a bidentate fashion, made this complex seem the most likely possibility. Furthermore, the enzyme seemed most able to stabilize the hydrated scissile peptide group in this orientation through a network of hydrogen bonds and electrostatic interactions (FIG. 6).

Our attention was now diverted back to the substrate data (TABLE 1) to determine if our modeled complexes are consistent with this data. Since the most drastic movement of our substrate upon proceeding from the Michaelis complex to the tetrahedral intermediate occurs at position P_2, we examined this data first. These substrates were reported[25] to follow Michaelis-Menten kinetics. Therefore, we made the assumption that the K_M values approximate the corresponding K_S values. Kinetic studies with a similar substrate by Fruton[29] support this assumption.

The standard state Gibbs energy change upon proceeding from free enzyme and substrate to the Michaelis complex is given by $\Delta G = -RT\ln\left[\frac{1}{K_S}\right]$.[24,30] The relationship $k_{cat}/K_M = kT/h \exp(-\Delta G^\ddagger/RT)$, derived from transition state theory, can be used to calculate the standard Gibbs energy of activation $(\Delta G^\ddagger)$ relative to the free enzyme and substrate.[24,30] The variable P_2 substrate data shown in TABLE 1 were converted into the reaction-coordinate diagram shown in FIGURE 8 via these equations.

Using the substrate with Z-Gly (G) at P_2 as a standard, we see that the P_2 phenylalanine side chain stabilizes both the Michaelis complex and the transition state, whereas only the latter is stabilized by a P_2 alanine side chain.

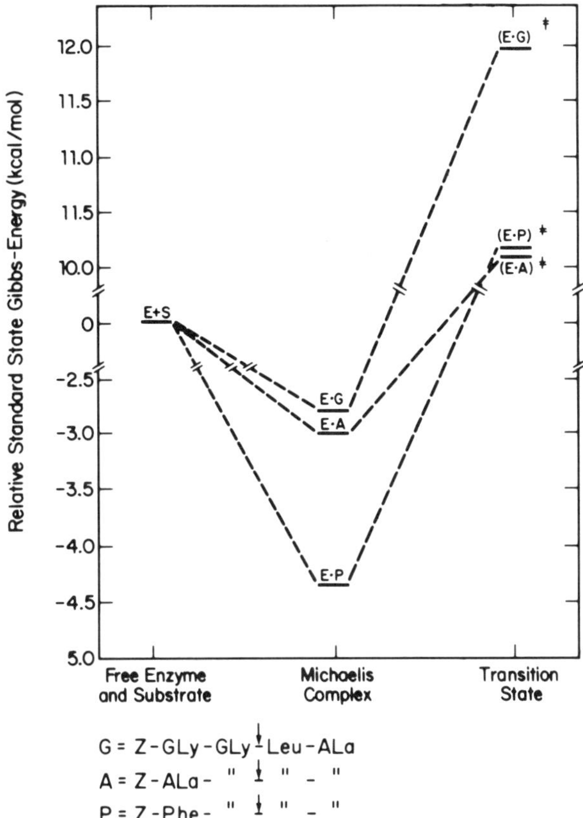

FIGURE 8. Reaction-coordinate diagram illustrating the standard state Gibbs energy changes for the conversion of thermolysin (E) and the variable P_2 substrates (S) listed in TABLE 1 to the Michaelis complex and then the transition state. The Gibbs energies were calculated at a standard state of IM.

We interpreted this to mean that the Phe phenyl group selectively stabilizes the Michaelis complex whereas the Ala methyl group (and the Phe methylene) selectively stabilize the transition state. The magnitude of these stabilizations are within the estimated range of intrinsic binding energy available from such groups.[26]

An examination of our preferred Michaelis complex and transition state suggested an explanation for this substrate data. We noted that, in the Michaelis complex (FIGURE 6), the Phe phenyl group was adjacent and perpendicular to the Tyr 157 aromatic ring. This arrangement is the one preferred for phenyl groups in proteins[31] and would be expected to stabilize the Michaelis complex. The Phe methylene, and by analogy an Ala methyl group, is also adjacent to the Tyr 157 but is closer to the hydroxyl group. This position may force an unfavorable desolvation of the hydroxyl group and an entropy loss by the substrate that is not significantly overcome by hydrophobic binding with part of the Tyr 157 aromatic ring. As a result, only the Phe phenyl group is suitably positioned to stabilize the Michaelis complex. In the tetrahedral intermediate (transition state) the Phe methylene (and Ala methyl group) is positioned above the His 146 side chain such that it may engage in hydrophobic binding and stabilize the complex. The Phe phenyl group can also engage in some hydrophobic binding but much of this group is positioned over an indentation in the bottom of the active site reducing the extent of such binding. Furthermore, it appears as though the Asp 150 side chain must rotate away from the active site and forfeit its hydrogen bond with Trp 115 in order to make room for this phenyl group. The balance of these interactions may be no net stabilization of the transition state by the Phe phenyl group.

We also examined the substrate data for positions P_1, P_1' and P_2' [25] in light of our Michaelis complex and tetrahedral intermediate complex. We found our complexes were consistent with this data as well.[12] Encouraged by these results, we continued to model the remaining steps in the reaction sequence, *i.e.,* bond cleavage and product release, using similar strategy, assumptions, and methods. A specific proposal for these subsequent steps resulted,[12] which is illustrated in FIGURE 6. These proposed steps are consistent with the available experimental data and do not require any significant motion of the active site residues (except Asp 150 as mentioned above).

The coordinates for the various modeled complexes generated in this study were used to construct an animation of the entire proposed mechanism. This animation was prepared by interpolating from one modeled complex to the next and displayed with GRAMPS. The animation further convinced us that the enzyme active site need not undergo any significant conformational changes during the course of the reaction.

Conclusions

The level of sophistication of the methods used in this study were adequate for our immediate needs, *i.e.,* to provide some theoretical background for our rational inhibitor design efforts. The results prompted us to propose[12] some key modifications and a number of extensions to the mechanism previously suggested.[18,22] Clearly, our understanding of this enzyme would be further enhanced if more rigorous methods were now applied. It would be interesting to determine if protein dynamics plays a role in the rate acceleration achieved by a relatively rigid enzyme such as thermolysin. For example, can the vibrational frequencies of the enzyme and substrate couple in a way that renders the scissile peptide linkage more labile?

We will report the results of our inhibitor design efforts in due course.

ACKNOWLEDGMENTS

We gratefully acknowledge Prof. B. Matthews for helpful discussions and crystallographic coordinates prior to publication. We also thank Dr. A. A. Patchett for encouragement and support during the course of these studies.

REFERENCES

1. GUND, P., J. D. ANDOSE, J. B. RHODES & G. M. SMITH. 1980. Science **208**: 1425.
2. GUND, P. & H. B. SCHLEGEL. 1981. Ann. N.Y. Acad. Sci. **367**: 510.
3. STILL, W. C. & D. GALYNKER. 1981. Tetrahedron **37**: 3981.
4. HUMBLET, D. & G. R. MARSHALL. 1981. Drug Devel. Res. **1**: 409.
5. BUSH, B. 1983. Comput. Chem. **8**: 1.
6. VEBER, D. F., R. M. FREIDINGER, D. S. PERLOW, W. J. PALEVEDA, JR., R. W. HOLLY, R. G. STRACHAN, R. F. NUTT, B. H. ARISON, C. HOMNICK, W. C. RANDALL, M. GLITZER, R. SAPERSTEIN & R. HIRSCHMANN. 1981. Nature **292**: 55.
7. BAUER, W., U. BRINER, W. DOEPFNER, R. HALLER, R. HUGUENIN, P. MARBACH, T. J. PETCHER & J. PLESS. 1982. Life Sciences **31**: 1133.
8. THORSETT, E. D., E. E. HARRIS, S. ASTER, E. R. PETERSON, D. TAUB, A. A. PATCHETT, E. H. ULM & T. C. VASSIL. 1983. Biochem. Biophys. Res. Comm. **111**: 166.
9. GUND, P. 1982. Trends Pharmacol. Sci. **3**: 56.
10. KUYPER, L. F., B. ROTH, D. P. BACCANARI, R. FERONE, C. R. BEDDELL, J. N. CHAMPNESS, D. K. STAMMERS, J. G. DANN, F. E. A. NORRINGTON, D. J. BAKER & P. J. GOODFORD 1982. J. Med. Chem. **25**: 1120.
11. BOGER, J. *et al.* 1983. Nature **303**: 81.
12. HANGAUER, D. G., A. F. MONZINGO & B. W. MATTHEWS. 1984. Biochemistry **23**: 5730–5741.
13. HOLMES, M. A. & B. W. MATTHEWS. 1982. J. Mol. Biol. **160**: 623–639.
14. ARGOS, P., R. M. GARAVITO, W. EVENTOFF & M. G. ROSSMANN. 1981. Biomolecular Structure, Conformation, Function and Evolution. Proc. Int. Symp. **1**: 205–225.
15. KESTER, W. R. & B. W. MATTHEWS. 1977. J. Biol. Chem. **252**: 7704–7710.
16. PATCHETT, A. A., E. HARRIS, E. TRISTRAM, M. WYVRATT *et al.* 1980. Nature **288**: 280–283.
17. MAYCOCK, A. L., D. M. DESOUSA, L. G. PAYNE, J. TEN BROEKE, M. T. WU & A. A. PATCHETT. 1981. Biochem. Biophys. Res. Commun. **102**: 963–969.
18. KESTER, W. R. & B. W. MATTHEWS. 1977. Biochemistry **16**: 2506–2516.
19. MONZINGO, A. F. & B. W. MATTHEWS. 1984. Biochemistry **23**: 5724–5729.
20. VOORDOUW, G. & R. S. ROCHE. 1975. Biochemistry **14**: 4667–4673.
21. HORROCKS, W. D. & D. R. SUDNICK. 1981. Acc. Chem. Res. **14**: 384–392.
22. HOLMES, M. A. & B. W. MATTHEWS. 1981. Biochemistry **20**: 6912–6920.
23. HOLMES, M. A., D. E. TRONRUD & B. W. MATTHEWS. 1983. Biochemistry **22**: 236–240.
24. FERSHT, A. 1977. *In* Enzyme Structure and Mechanism. W. H. Freeman. San Francisco.
25. MORIHARA, K. & H. TSUZUKI. 1970. Eur. J. Biochem. **15**: 374–380.
26. JENCKS, W. P. 1981. Proc. Nat. Acad. Sci. U.S.A. **78**: 4046–4050.
27. HAMMOND, J. 1955. J. Am. Chem. Soc. **77**: 334–338.
28. SLOAN, D. L., J. M. YOUNG & A. S. MILDVAN. 1975. Biochemistry **14**: 1998–2008.
29. MORGAN, G. & J. S. FRUTON. 1978. Biochemistry **17**: 3562–3568.
30. PAGE, M. I. 1980. Int. J. Biochem. **11**: 331–335.
31. THOMAS, K. A., E. M. FLUDER & G. M. SMITH. Unpublished results. (See also Thomas, K. A., G. M. Smith, J. B. Thomas & R. J. Feldmann. 1982. Proc. Nat. Acad. Sci. U.S.A. **79**: 4843–4847.)

Design of Anticancer Drugs Using Modeling Techniques

V. N. BALAJI,[a] J. SCOTT DIXON, D. H. SMITH,
R. VENKATARAGHAVAN, AND K. C. MURDOCK

Lederle Laboratories
Pearl River, New York 10965

INTRODUCTION

Several classes of anticancer drugs are hypothesized to exert their activity through interaction with DNA.[1,2] Binding of anticancer drugs to nucleic acids has been widely investigated and two types of structural models have been proposed to describe the interactions. The first involves intercalation of a drug into double-helical DNA.[3] The second model involves nonintercalative binding.[4] In this investigation we have focused on intercalative binding in order to investigate drug-DNA interactions as a first step in the rational design of new anticancer drugs.

There are several important factors to be considered when approaching the question of design. In this paper we put aside factors related to drug absorption, transport and metabolism and focus on the structural aspects of the actual drug-DNA interactions. Here there are two important considerations that we address directly. The first is the determination of favorable DNA intercalation geometries. The second is the mode(s) of interaction of representative drug molecules with these geometries to produce the intercalated drug-DNA complexes.

Modeling Studies

The determination of intercalation geometries and the investigation of drug-DNA interactions have been undertaken using the methods of X-ray crystallography and computer-based molecular models derived in part from the crystallographic results. We discuss in the next two sections results obtained from these studies that are pertinent to our work.

X-Ray Crystallography

The conformational features of the intercalation sites in polynucleotides have been reviewed.[5] More recent results, including summary of the geome-

[a] Present address: Allergan Pharmaceuticals, 2525 Dupont Drive, Irvine, CA 92715.

tries of dimer duplexes of DNA fragments intercalated with typical drugs, have been presented.[6,7] These studies showed that base pair separations at the intercalation site were about 6.8 Å, accompanied by an unwinding of the base pairs from the classical B-DNA value of 36°. Alternate sugar puckering was not required for intercalation. However, intercalation geometries in polynucleotides required conformational changes in adjacent nucleotides. The base-turn angle was found less sensitive to the conformation of the backbone than to small alterations in the base-pairing geometry. It was also postulated that this angle was dependent on the nature of the intercalating drug.

The crystallographic study of daunomycin-hexanucleoside complex[8] raises other interesting possibilities, in particular, about the assumption of unwinding concentrated at the intercalation site itself. In this structure, at the intercalation site, there seems to be no unwinding at all and the unwinding of base pairs occurs in the base pairs adjacent to the intercalation site.

These studies suggest that computational approaches to modeling drug-DNA intercalation should consider, for completeness, modifications to base pair and backbone geometries beyond the intercalation site itself. In this regard, the crystal structure of B-DNA dodecamer[9] indicates flexibility of base-base twist parameters ranging between 27° to 40°, a further illustration of the inherent flexibility of the double helix.

Molecular Modeling

The generation of intercalation geometries starting from internal parameters of bond lengths, bond angles, and torsion angles has been attempted by several groups.[10–18] The generation of such geometries is based on crystallographic data such as those summarized above. Therefore, computer-based modeling studies generally assume that at the intercalation site the bases are separated by approximately 6.8 Å, associated with an unwinding from the standard B-DNA twist of 36°. The perturbation of the overall geometry is generally confined to the intercalation site or extended to include base pairs adjacent to the site. The conformations next to the intercalation site are normally forced to the B-DNA geometry. For example, the proflavin intercalation geometry was simulated with the helix axis of a polynucleotide constrained to be linear, with the base pairs stretched apart at the intercalation site from approximately 3.4–6.8 Å separation.[10,11] This change was accomplished by rotational changes around the -C5'-C4'- and -P-O5'- bonds.[b] These changes were accompanied by smaller conformational changes at the adjacent base pairs. In the derived B-DNA model there is alteration of sugar pucker at the intercalation site. In the corresponding A-DNA model all the sugars retain their C3'-endo pucker.

An algorithm has been reported to determine the backbone conformation of nucleic acids,[12] and intercalation geometries have been generated for

---

[b] IUPAC-IUB nomenclature[19] for the designation of atoms in polynucleotides and of torsion angles is followed throughout this paper.

a tetramer duplex extracted from B-DNA.[13] These tetranucleotide duplex models accommodate conformational variations in three successive dimer units and allow as well for dislocation of the helix axes on either side of the binding sites.

A series of structures of double-stranded DNA involving an intercalation site has been reported.[14] Backbone conformations were solved for parallel base pairs at different separations.

Helical parameter analysis of the dinucleotide units with intercalation sites in general show mixed sugar puckerings.[15] Studies of drug intercalation into left-handed Z-DNA[15,16] and B-DNA[16] have been reported. Model building studies of stacked structures of dinucleotide systems have been analyzed. For a series of stacked family of structures a corresponding intercalation form was found to exist.[17] Recently, Taylor and Olson[18] have carried out a combined geometric and potential energy analysis to identify intercalation geometries. They found several conformations with unusual sugar puckering combinations and various phosphodiester arrangements. A large proportion of the energetically favored intercalation sites were closely related to the backbone conformation of familiar double-helical models such as A-, B-, Z-, and Watson-Crick DNA.

The above studies all have relevance to the generation of intercalation geometries and were considered in this investigation. In particular, we were interested in evaluating the flexibility and relative energies of alternative intercalation sites and in exploring the ranges of twist angles and helical axis shifts corresponding to low-energy sites.

Drug Molecules

Compounds that have been shown to bind to DNA by intercalation[6] have generally been condensed tricyclic aromatic ring systems with at least one basic function. Many of these compounds are active as antitumor agents.[6] In this study we have directed our attention to two representative structures of this type, both of which are currently undergoing clinical trials as anticancer agents. These compounds are 1,4-dihydroxy-5,8-bis[[2-[(2-hydroxyethyl)amino]-ethyl]amino]-9, 10-anthracenedione (mitoxantrone)[20] and 9,10-anthracenedicarboxaldehyde bis[(4,5-dihydro-1H-imidazoyl-2-yl)hydrazone] (bisantrene).[21] Mitoxantrone and bisantrene display potent anticancer activities in animal models against leukemias and certain solid tumors.[20,21] Both compounds are significantly less cardiotoxic than other anticancer drugs such as adriamycin.[20,21] We have considered these compounds as representative anticancer drugs to study possible mechanisms of interaction with different intercalation sites in DNA to determine features of the drug-DNA interactions that will be relevant for design of new drugs.

Several additional studies have bearing on the modes of intercalation of these compounds. Physicochemical studies on DNA binding specificity and RNA polymerase inhibitory activity of bis(aminoalkyl)anthraquinones (mitoxantrone and an analog) and bis(methylthio)vinylquinolinium iodides have been reported.[22] A rough correlation between anticancer activity and DNA binding ability of anthraquinones and quinolium compounds was apparent. Binding studies indicated two binding sites on DNA for mitoxantrone. The stronger binding was presumed to be intercalation between successive base pairs. The weaker binding was considered to be an electrostatic interaction between DNA phosphate groups and the amino side chains of mitoxantrone. Also, comparison of absorption spectra and the extent of RNA transcription inhibition in systems possessing a range of ratios of DNA base pairs showed preferential binding involving GC base pairs.

An electron microscopic study of the interaction of mitoxantrone and bisantrene with DNA[23] has revealed evidence for intercalative binding of both molecules. The former causes a 13% average length increase in pBR322 corresponding to approximately 580 drug molecules per circle at saturation whereas the latter causes an 11% increase in length corresponding to approximately 480 drug molecules bound per circle. Consideration of the known GC preference for binding of the drugs and inspection of the known sequence of pBR322 suggested that the available intercalation sites were occupied and that additional external electrostatic binding of the cationic drugs also occurs. Mitoxantrone caused a 16.6% length increase in nicked PM2, which is equivalent to approximately 1700 drug molecules per circle. Electron microscopic measurements on relaxed PM2 with progressively increasing proportions of mitoxantrone, from 1.4:1 to 14:1 drug molecules per base pair, revealed the onset of formation of lace-like networks of DNA circles linked together. This phenomenon, which is not produced by bisantrene, was attributed to the formation of inter-DNA links by the charged side arms of mitoxantrone and is in accord with previous reports that mitoxantrone causes severe compaction and distortion of chromatin.[24,25]

In light of this evidence, we have explored drug-DNA interactions in representative systems using computed intercalation geometries together with molecular models of mitoxantrone and bisantrene. The details of our methods and results are described in the following sections.

METHODS

Our methods include the following major steps:

1. Determination of intercalation geometries. We first examined low-energy conformations of alternative geometries for the intercalation site itself.
2. Modeling of intercalants. We next derived three-dimensional models for our representative drugs, mitoxantrone and bisantrene.
3. Drug-DNA interactions. We used conformational energy calculations to investigate the details of intercalation of the drug models into the derived intercalation geometries.
4. Presentation and evaluation of results. We used a variety of techniques of computer graphics to visualize the geometries of the independent

and the intercalated systems, and to examine the electrostatic and hydrogen bonding aspects of the drug-DNA interactions.

Before discussing these major steps, we summarize some of the general methods used in our calculations.

Conformational Energy Parameters

The conformational energies we compute are comprised of London dispersion, van der Waals repulsive, coulombic electrostatic, induced dipole, intrinsic torsional, hydrogen bond, bond angle bending and bond length distortion terms.[26-28] Also included were *gauche* effects of phosphodiester oxygen lone-pair orbitals,[29] electrostatic screening of phosphates and counterion condensation effects.[17] A potential function that uses partitioned torsional parameters[30] was used for the sugar moieties.

Charges

The atomic charge densities of mitoxantrone and bisantrene were computed by the quantum chemical method MINDO/3.[31] The charges on DNA fragments were determined in the same way. The effects of counterions associated with the pendant phosphate oxygens were taken into account by a correction of atomic charge.[32] Additional details concerning these terms and their associated parameters will be published elsewhere.

Intercalation Geometries

It is computationally prohibitive to consider derivation of intercalation geometries for extended polynucleotide systems while allowing variation in all degrees of freedom possible in such systems. Therefore, several simplifying assumptions were made to derive these geometries, as summarized below.

We have considered as our model double helical nucleotides a tetrameric unit whose relevant geometrical parameters are shown in FIGURE 1. The base pair geometries, bond lengths and bond angles were taken from Arnot and co-workers.[33-36] We have considered geometrical variations in the intercalation site itself and in the base pairs immediately adjacent to the site. Referring to FIGURE 1, a right-handed Cartesian coordinate system was chosen such that y-axis is the pseudodyad axis, and the z-axis is perpendicular to the base pairs. The base pairs were fixed initially at $dy = \sqrt{(r^2 - d^2)}$ where $r = r(\text{Cl}') = 5.86$ Å and $2d = (\text{Cl}'\text{-Cl}') = 10.85$ Å. The four bases bp1, bp2, bp3 and bp4 were symmetrically displaced about the pseudodyad axis by twist angles of $(t_2/2 + t_1)$, $(t_2/2)$, $(-t_2/2)$ and $(-t_2/2 - t_3)$, respectively. The twist angle at the intercalation site is t_2; t_1 and $t_3 = 36°$ are the twist angles corresponding to native B-DNA. The z coordinates for the base pairs were established at $(d_2/2 + d_1)$, $(d_2/2)$, $(-d_2/2)$ and $(-d_2/2 - d_3)$, respectively where $d_2 = 6.76$ Å, the base separation at the intercalation site, and $d_1 = d_3 =$

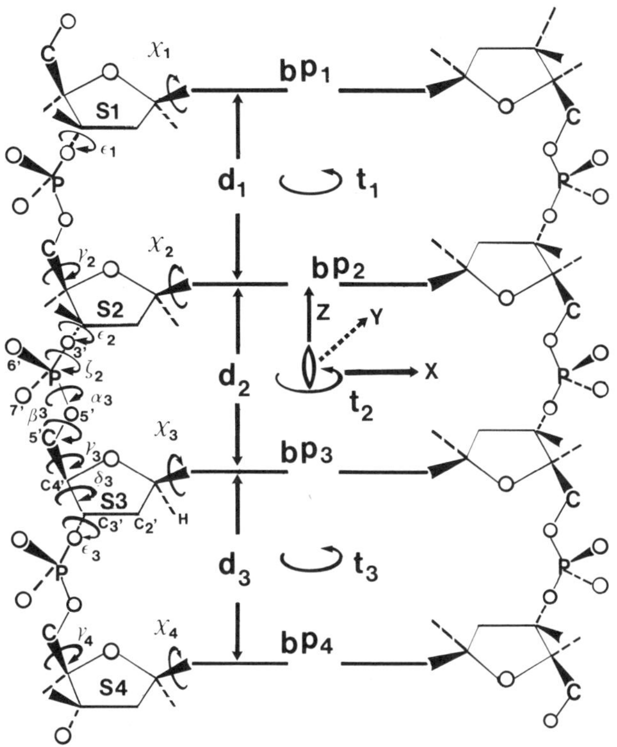

FIGURE 1. A representation of the DNA tetramer duplex used to explore potential intercalation geometries. Twist angles corresponding to the B-DNA geometry are indicated by t_1 and t_3, and t_2 is the twist angle at the intercalation site. The base pair separations for B-DNA correspond to d_1 and d_3; d_2 corresponds to the base separation for the intercalation site.

3.38 Å, the base separation for native DNA. Sugar moieties were generated using pseudorotation parameters ξ_1 and ξ_2.[37] The sugars were fixed using glycosidic torsional parameters χ_1, χ_2, χ_3 and χ_4. Initially χ_1 and χ_4 were fixed at the standard B-DNA geometry. The sugars S1 and S4 were fixed at C2′-endo geometries. To start with, the sugars S2 and S3 were generated corresponding to different combinations of C2′-endo and C3′-endo geometries. Initial values chosen for the relevant torsional angles are summarized in TABLE 1.

Conformational energy calculations have shown that the torsion angle about the C3′-O3′ bond (ε) can assume a range of values between 150° to 300° and 150° to 260° when the sugar geometries lie in the C2′-endo and the C3′-endo region, respectively, of the pseudorotational space.[38] Therefore, the different starting values of ε_3 and ε_2 chosen were 150°, 180°, 210°, 240° and 290°. All possible starting combinations of the above angles were used in preliminary calculations (energy minimization) using a coarse grid for variation of the torsional parameters. These preliminary calculations resulted

TABLE 1. Initial Values for Torsion Angles

Torsion angle	Starting values (in degrees)
γ_4	$+60$
γ_2	$-60, +60, 180$
γ_3	$-60, +60, 180$
ε_1	-160
χ_1	-120
χ_2	$-150, -120$
χ_3	$-150, -120$
χ_4	-120

in intercalation geometries wherein many of the above angles, independent of starting point, converged to singular values, except for ε_2 and γ_3. These latter angles were observed to converge in the ranges of $\varepsilon_2 = -60°$ to $180°$ and $\gamma_3 = 80°$ to $140°$ and $-80°$ to $-140°$. The preferred sugar geometries for S2 and S3 were C3'-endo and C2'-endo, respectively. Therefore, in subsequent minimizations, the starting values for γ_4, ε_3, γ_2, ε_1, χ_2, χ_2, χ_3, and χ_4 were initialized to these values, namely, $40°$, $-140°$, $-120°$, $-140°$, $-120°$, $-160°$, $-120°$, and $-120°$, respectively. The sugar conformations of S2 and S3 were initialized as above, and starting values for the angles ε_2 and γ_3 were chosen from the ranges listed above.

For the remaining possible starting conformations, minimization of conformational energy was carried out in the twist-shift plane where shift corresponds to the displacement of the B-DNA helix axis in the x-direction on either side of intercalation site, as shown in FIGURE 2. In practice, the initial few iterations of the minimization were done with frozen sugar conformations; these were later relaxed. A statistical weight of 1.0 was assigned to all the parameters varied except for γ_4, ε_1, χ_1 and χ_4 for which the weight of 0.2 was chosen when updating the variation of parameters in these angles for every iteration. This initial weighting scheme was used to seek solutions such that the base pairs adjacent to the intercalation site are near to the B-DNA geometry.

Sequence Specificity

We wished to examine intercalation sites comprised of different sets of base pairs. Previous studies have shown that the sugar-phosphate backbone geometry shows no correlation to the actual base sequence, for example, in the B-DNA crystal structure of the dodecamer d(CGCGAATTCGCG).[39] Therefore, our models of various tetramer duplexes (below) were constructed using the minimum energy geometries, derived as above, and substituting alternative base pairs into the backbone geometries. Conformational energies were, however, recalculated, but without further minimization.

In this investigation we have focused our attention on different interca-

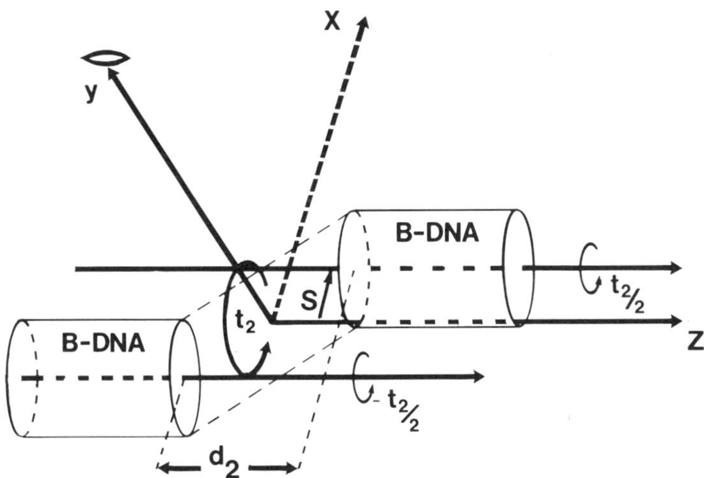

FIGURE 2. A representation of B-DNA with a twofold intercalation site. Base pair separation at the intercalation site is d_2. On either side of the site the helix is rotated by $+t_2/2$ and $-t_2/2$, where t_2 is the twist angle, and shifted by $+s$ and $-s$ along the x-axis.

lation geometries that possess a twofold axis of symmetry. The base sequences considered at the intercalation site included the following, considering both possibilities of chain direction pyr(5'-3')pur and pur(5'-3')pyr: (1) d(CpG) · d(GpC), (2) d(GpC) · d(CpG), (3) d(ApT) · d(TpA), and (4) d(TpA) · d (ApT).

Modeling of Mitoxantrone and Bisantrene

Molecular models of mitoxantrone (FIG. 3) and bisantrene (FIG. 4) were derived from structural subunits retrieved from the Cambridge Crystal File.[40]

Mitoxantrone. Mitoxantrone consists of a tricyclic anthraquinone moiety and two symmetrical side chains. At physiological pH the side chain nitrogen N(2) will be protonated, leading to NH_2^+ groups. The tricyclic ring system and the associated hydroxyl groups of mitoxantrone were fixed as found in the crystal structure of 1,4-dihydroxyanthraquinone.[41] The hydroxyl groups (O1-H2) are involved in intramolecular hydrogen bonds with the quinone carbonyl groups. The torsional parameters for a complete description of the side chain are given as θ_1-θ_7, as shown in mitoxantrone. These torsion angles were varied in subsequent calculations on the drug-DNA complex (see below). The side chain nitrogen atoms N(1) were kept planar as found in similar crystal structure moieties. The N(1)-H(4) groups are also involved in hydrogen bonds with the quinone carbonyl groups. Thus, θ_1 was held fixed at 180°.

Bisantrene. Bisantrene is comprised of a tricyclic anthracene moiety possessing two symmetric side chains attached to the central ring. The anthra-

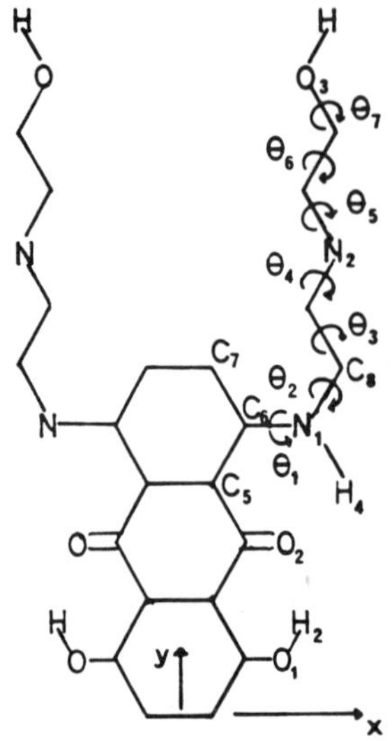

FIGURE 3. The structure of mitoxantrone, including designations for side chain atoms and torsion angles. We define θ_1 as [C(5)-C(6)-N(1)-C(8)].

cene ring system was fixed in a planar configuration. The side chain conformations are described by the conformational angles θ_1-θ_4, as shown in bisantrene. These angles were varied in subsequent calculation of the drug-DNA complex (see below).

Modeling of the Drug-DNA Complex

Models of intercalation of anthracycline analogs into DNA have been proposed previously.[42-46] In these models, usually the long axis of the tricy-

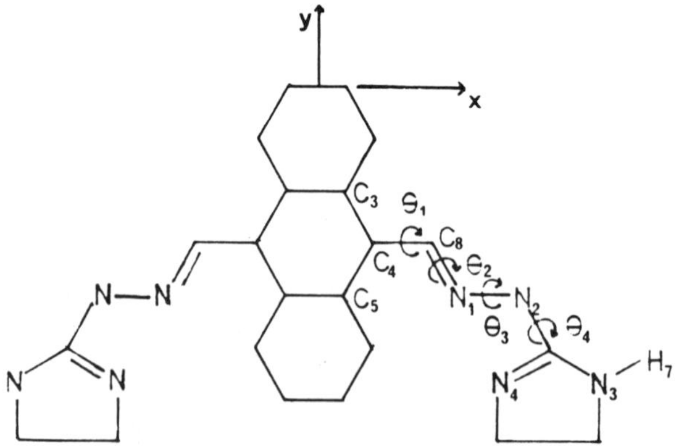

FIGURE 4. The structure of bisantrene, including designations for side chain atoms and torsion angles. We define θ_1 and θ_4 as [C(3)-C(4)-C(8)-N(1)] and [N(1)-N(2)-C-N(3)], respectively.

clic ring system is aligned parallel to the long axis of the base pairs and is inserted from the major groove.[42-45] Alternatively, minor groove intercalation has been proposed to occur with the long axis of the drug aligned perpendicular to the long axis of the base pairs.[46] In a first step of determination of our drug-DNA complexes, we examined the steric feasibility of fitting of mitoxantrone and bisantrene in a representative DNA intercalation geometry using interactive computer graphics (see below). This examination was done to determine the possibility for intercalation from the major and/or minor grooves of the double helix. Also, this examination revealed that side chains of mitoxantrone and bisantrene could interact with backbone atoms at the intercalation site, but were too short to interact with backbone atoms of the adjacent base pairs.

The twofold axes of mitoxantrone and bisantrene and the dyad axis of the intercalation site (FIG. 1) were chosen to be the y-axis for convenience. The tricyclic ring systems of mitoxantrone and bisantrene were fixed in the xy-plane. In preliminary calculations we investigated the conformational energy of intercalation as a function of penetration depth along the y-axis. In these calculations, we did not consider the entire side chains of mitoxantrone and bisantrene. For mitoxantrone, the side chains beyond C9 were removed, C9 was replaced by a methyl group and θ_1 and θ_2 were set to 180°. For bisantrene, the side chains beyond N(2) were removed, N(2) was replaced by a methyl group and θ_1 and θ_2 were set to 90° and 180°, respectively.

Given the approximate intercalation geometries of the drug-DNA complex, we completed the calculations by energy minimization of the entire complex, now including drug side chains. We made no attempt to explore exhaustively all possible starting points for the flexible side chains. Rather, we calculated possible geometries and selected as starting points only representative complexes that positioned side chain atoms near the helix backbone. The selection was done using calculations to eliminate complexes with bad contacts (van der Waals overlaps) and computer graphics to choose among the remaining geometries. Thus, although our calculations are not exhaustive, the geometries chosen and the results obtained from them are sufficient to support our major findings.

Computer Graphics

Interactive computer graphics were essential for preliminary definition of intercalation geometries and examination of results of the calculations. The graphics system employed was an Evans and Sutherland Multi-Picture System driven by a Digital Equipment Corporation VAX-11/780. Major software systems used to visualize and manipulate the structures included GRAMPS[47] and FRODO.[48]

Color-coded computer graphics representations of electrostatic potentials have been used to determine electrostatic and topographic complementarity in macromolecule-ligand interactions.[49] Such electrostatic potential surfaces were used in several steps of our investigations to examine the complementarity of mitoxantrone and bisantrene with the DNA intercalation geometries. We first generated molecular surfaces for the drug and tetramer duplex.[50] We computed the electrostatic potential V at point r for a system of charges $q(i)$ at points $r(i)$ in a medium of dielectric constant ε, using the following equation:

$$V(r) = \sum \frac{q(i)}{\varepsilon|r(i) - r|} \cdot \tag{1}$$

The electrostatic potential was calculated 1.4 Å along the surface normal vector from a given molecular surface point and represented at the surface point itself. Atoms within 20 Å were considered for construction of the potential at point r.

TABLE 2. Selected Starting and Corresponding Minimum Energy Conformations for the Intercalation Geometry for the DNA Tetrameric Unit pdApdTpdApdT

Starting conformation (degrees)		Minimum energy conformation (degrees)		Energy[a] (kcal/mole)	t_2 (deg)	s (Å)	Remarks
γ_3	ε_2	γ_3	ε_2				
80	−60	95	−74	−72.4	20	1.0	
80	−90	93	−78	−73.5	20	1.0	
80	−120	91	−105	−70.7	20	1.0	
80	−150	107	−147	−71.3	22	0.7	case 3
80	180	81	−167	−63.7	30	0.5	
110	−60	100	−74	−73.2	20	1.0	
110	−90	95	−89	−74.0	20	0.9	case 4
110	−120	96	109	−69.3	20	0.5	
110	−150	111	−148	−71.0	20	0.5	
110	180	125	−172	−71.3	28	0.0	case 2
140	−60	126	−60	−60.0	15	1.0	
140	−90	155	−85	−49.9	25	0.0	
140	−120	126	−134	−59.3	25	0.0	
140	−150	133	−142	−58.0	25	0.0	
140	180	135	−171	−73.2	30	0.5	
−80	−60	−78	−74	−63.2	0	−0.5	
−80	−90	−61	−102	−72.4	2	−1.1	case 5
−80	−120	−65	−105	−67.4	0	−1.0	
−80	−150	−88	−147	−68.6	5	−1.5	
−80	180	−89	−169	−65.9	10	−1.5	
−110	−60	−101	−49	−59.2	5	0.0	
−110	−90	−95	−102	−51.4	5	−1.0	
−110	−120	−95	−132	−64.7	5	−1.5	
−110	−150	−82	−150	−71.4	8	−1.5	case 1
−110	180	−97	−171	−65.3	10	−1.5	
−140	−60	−150	−54	−60.6	−5	0.0	
−140	−90	−144	−75	−55.5	−5	−0.5	
−140	−120	−125	−134	−47.7	0	−1.5	
−140	−150	−125	−159	−59.2	0	−1.5	
−140	180	−127	−194	−60.9	0	−1.5	

[a] Energy/strand of the tetramer duplex.

RESULTS AND DISCUSSION

The hypothesis that planar drug molecules intercalate between base pairs of a DNA helix is well established. The planar moiety of the drug and the base pairs between which the drug is sandwiched are all parallel and the perpendicular distance between the base pairs is approximately 6.76 Å. Another crucial geometrical parameter of interest is the twist of base pairs at the intercalation site. If we allow no shift of the helix axis at the intercalation site, local energy minima for the intercalation geometries are found at twist angles in regions centered at 8°, 20° and 28° independent of the drugs intercalated. These results are in substantial agreement with those of Miller and Pycior.[13] However, if a shift in the helix axis perpendicular to the pseudodyad axis of the intercalation site is allowed, the accessible twist angle range can vary from 0° to 30°, as shown below.

The flexibility of the intercalation sites was examined by considering the twist angles at the intercalation site together with shifts in the B-DNA helix axis on either side of the intercalation site along the *x*-axis (see FIG. 2). Plots of the allowed region in the twist-shift plane were obtained for different starting points of the sterically feasible first-order combinations of backbone geometries of the intercalation site.

Previously (see METHODS) we described preliminary calculations that allowed us to fix several of the backbone torsional angles (FIG. 1), and to establish the sugar puckerings of S2 and S3 as C3′-endo and C2′-endo, respectively. Initially, for the thirty (6 × 5) starting combinations of γ_3 (80°, 110°, 140°, −80°, −110°, −140°) and ε_2 (−60°, −90°, −120°, −150°, 180°), the energy of the tetramer duplex of DNA was minimized in the twist-shift plane in the range of shift $s = -1.5$ Å to $+1.5$ Å at 0.5 Å increments and twist in the range of $t_2 = -5°$ to $+40°$ at increments of five degrees (TABLE 2). The minimum energy regions obtained at this coarse interval of twist and shift were further examined at a finer interval of one degree in twist and 0.2 Å in shift. Data on selected minimum energy conformations thus obtained are given case numbers in the remarks column of TABLE 2. This table does not present values for sugar conformations and the glycosidic torsion angles, which were within five degrees from the starting conformations. Features of the energy contour maps in the twist-shift (t,s) plane corresponding to two representative starting combinations are described below. These conformations are those found to be favorable for intercalation of mitoxantrone and bisantrene, respectively.

- *Case 1.* The starting combination of $\gamma_3 = -110°$ and $\varepsilon_2 = -150°$ (which minimizes to −82°, −150°, respectively), yields the (t,s) map shown in FIGURE 5. The minimum energy (−71.4 kcal/mole) conformation occurs at twist = 8°, shift = −1.5 Å. This conformation is approximately that shown in FIGURE 6 (see below). The 5 kcal/mole contour above the minimum encompasses a shift range of −1.8 to −1.2 Å and a twist range of −3° to +10°.

- *Case 2.* The starting combination of $\gamma_3 = 110°$, $\varepsilon_2 = 180°$ (which minimizes to 125°, −172°, respectively) leads to the (t,s) map shown in

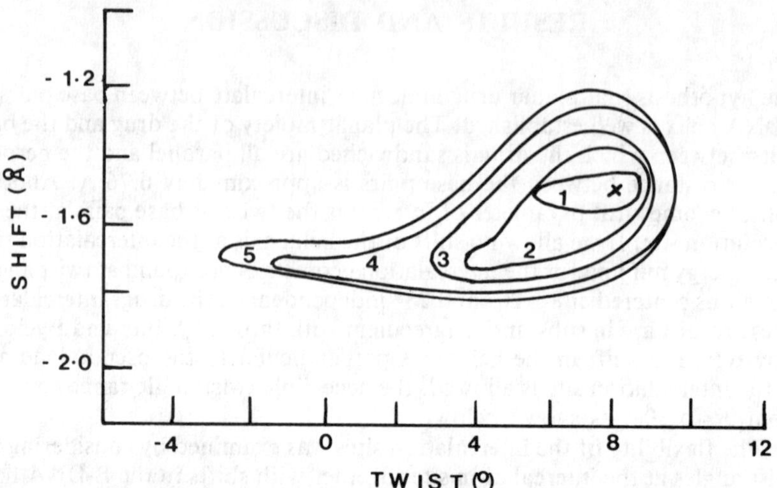

FIGURE 5. Conformational energy map in the twist-shift plane for case 1 (see text for explanation). Isoenergy contours are in kcal/mole references to the minimum, point x, taken as zero.

FIGURE 7. The minimum energy (-71.3 kcal/mole) occurs at 28° degrees and about 0.0 Å. The minimum energy conformation is shown in FIGURE 8. The 5 kcal/mole contour above the local minimum encloses a sharp region in shift of 0.0 to -0.2 Å with a twist range from 22° to 31°.

Depending on the starting combination of the main chain torsional parameters, the twist-shift plane showed several regions of favorable intercalation geometries. These geometries were examined using computer graphics to understand the phosphate backbone orientation which in turn is important for examining the specific interaction of drug molecules. The drugs Mitoxantrone and bisantrene with their side chains in the extended conformation were examined with several intercalation geometries, including those mentioned above, to explore the possibility of hydrogen bonds involving the NH or OH groups with phosphate oxygens of the DNA backbone.

DNA-Mitoxantrone Intercalation

Our preliminary investigations revealed that mitoxantrone can intercalate from either the minor or major grooves of the DNA intercalation site. More detailed studies indicated that the planar ring system of mitoxantrone can intercalate in either groove such that the principal long axis of the ring system is perpendicular to the DNA helix axis and parallel to the dyad axis of the intercalation site. However, once the side chains were included in the investigation, it became apparent that the interaction is more favorable from the minor groove than the major groove. The depth to which the ring system can penetrate the intercalation site is restricted by (a) the interaction between

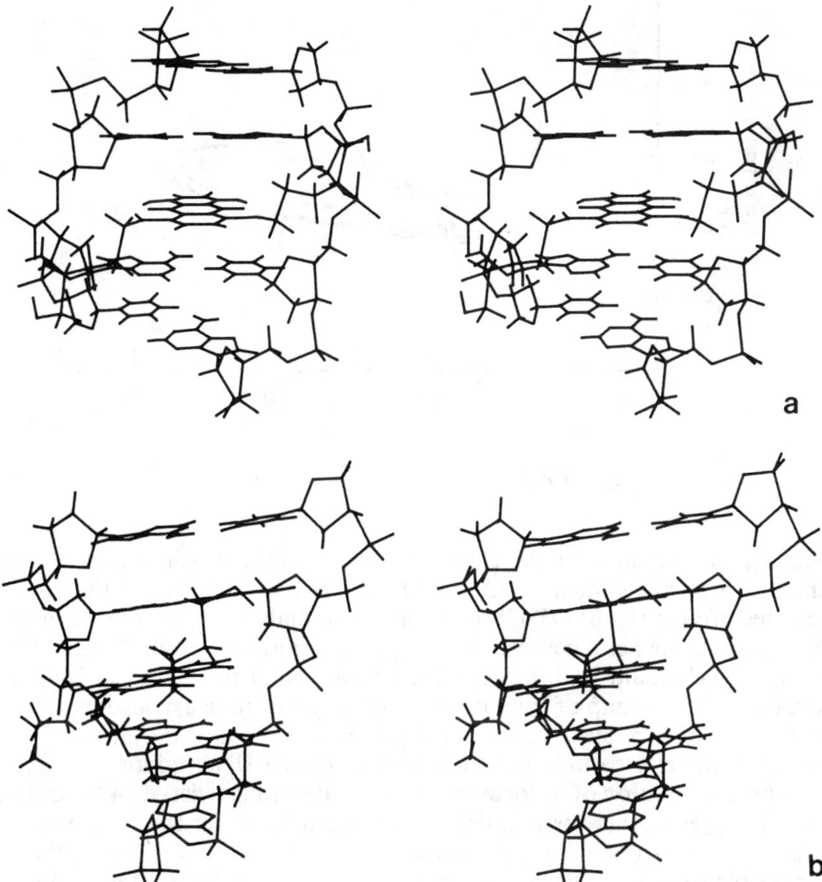

FIGURE 6. **(a)** Stereoview of the intercalation of mitoxantrone into a typical intercalation site (see text for details), looking down the minor groove. The helix axis is tilted towards the observer by 10° to illustrate the stacking of the tricyclic moiety between the base pairs. **(b)** Same as in **(a)**, but rotated 45° around the helix axis, showing the hydrogen bonding features of the side chain hydroxyl groups of mitoxantrone with phosphate oxygens of the backbone leading to cross-linking of the complementary strands.

the ring system and the nearest neighbor base pairs and (b) the interaction between the side chains of mitoxantrone and the sugar phosphate groups of the DNA duplex. Both types of interaction are more favorable for mitoxantrone when intercalation occurs from the minor groove. Our observations also revealed that the N(1)-H(4) group is unavailable for hydrogen bonding either with the bases or the phosphate oxygens of the backbone.

Interactions were first examined by computing the electrostatic potential surfaces of mitoxantrone and representative tetramer duplexes. From the examination of these surfaces we observed several important features. There

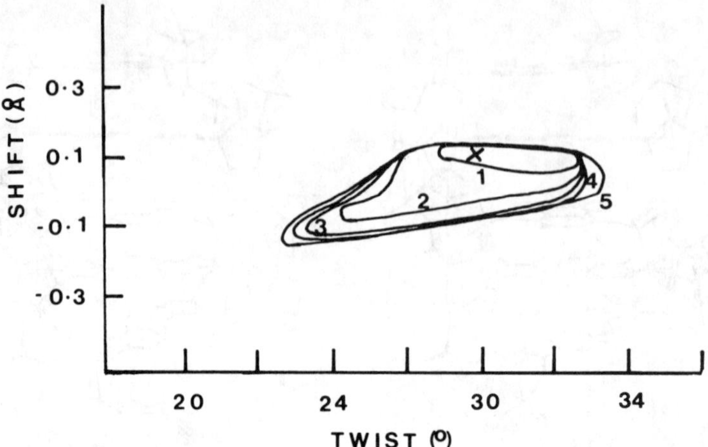

FIGURE 7. Same as in FIGURE 5, for case 2.

was a good matching of the electrostatic potential between the base pairs and the three-ring system. DNA has a negative electrostatic potential pocket centered around the -P-O5′-C5′-position of the chain. Further, the side chain of mitoxantrone has a complementary positive potential around the $N(2)H_2^+$ groups which could lead to favorable interaction at this position. The terminal hydroxyl group of the side chain of mitoxantrone can also hydrogen bond with a phosphate oxygen when the side chain of mitoxantrone is aligned to match the electrostatic potential surface of the DNA minor groove.

The intercalation of mitoxantrone into a site with a twist of 24° and shift of 1 Å suggests other possibilities for hydrogen bonding. In this geometry, which is 4 kcal/mole above the nearest local minimum, we observe that it is possible for the $N(2)H_2^+$ groups in the side chains to hydrogen bond with the phosphate oxygens that are oriented towards the minor groove. In this case the terminal hydroxyl groups are free to hydrogen bond with the phosphate oxygens of another chain of DNA, which would lead to cross-linking of DNA chains.[24,25] We expect that there are other intercalation geometries wherein such interhelical cross-links are possible.

We next examined the energy vs. depth profiles of intercalation of moieties (see METHODS) of mitoxantrone (and bisantrene) into representative, low-energy intercalation geometries, e.g., cases 1 and 2 above. The results for five such sites are summarized in TABLE 3. In the case of mitoxantrone, intercalation from the minor groove side of the tetramer duplex is favored over major groove intercalation by 2–8 kcal/mole. Only one of the sites, case 2, showed a preference for major groove intercalation. A typical energy vs. depth profile for minor groove intercalation of mitoxantrone is shown in FIGURE 9, based on case 1, the minimum energy geometry used in these calculations.

The final step in our calculations is to consider the details of the energetics of interaction of mitoxantrone with complete side chains with relaxed

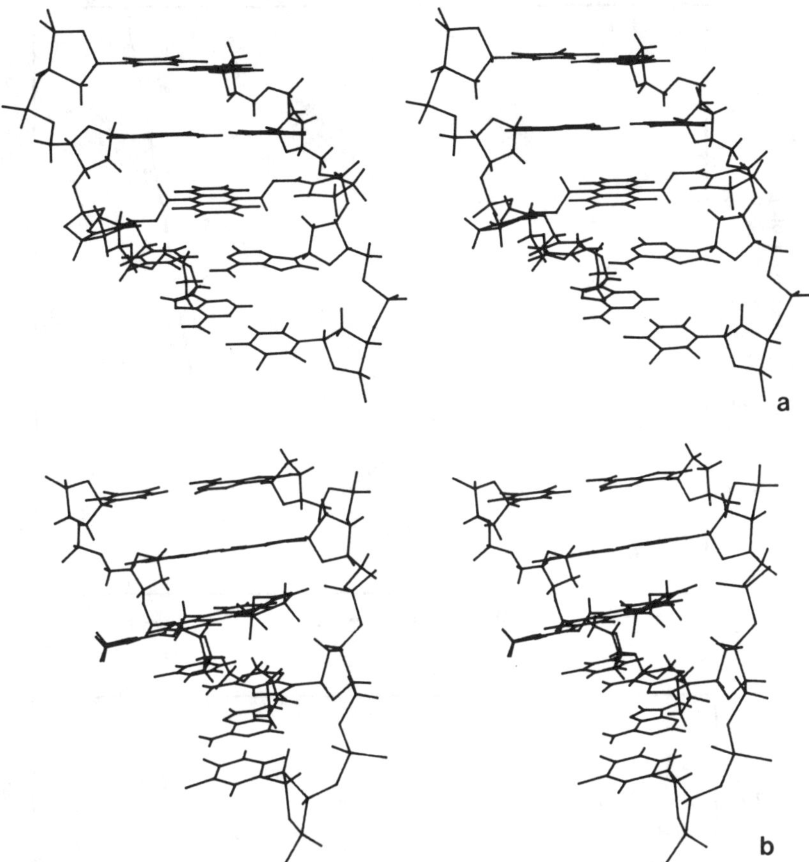

FIGURE 8. (a) Stereoview of the intercalation of bisantrene into a typical intercalation site (see text for details) looking down the major groove. The helix axis is tilted towards the observer by 10° to illustrate the stacking of the planar moiety of bisantrene between the base pairs. **(b)** Same as in **(a)**, but rotated 45° around the helix axis, showing the hydrogen bonds between dihydroimidazole NH groups with the phosphate oxygens of the backbone cross-linking the complementary strands.

tetramer duplex geometries. Given favorable values for intercalation depth, we are now investigating possible side chain conformations of mitoxantrone by examining allowed (*trans, gauche +, gauche −*) combinations of the torsion angles θ_2-θ_6. The side chains are being examined for favorable interaction with the sugar-phosphate backbone of DNA. The results of these studies will be reported in a subsequent publication.

As an example of one of the drug-DNA complexes obtained by our methods, we show the intercalation of mitoxantrone into the geometry represented by case 1 in FIGURE 6. The stereo pair in FIGURE 6b illustrates a side view of the complex.

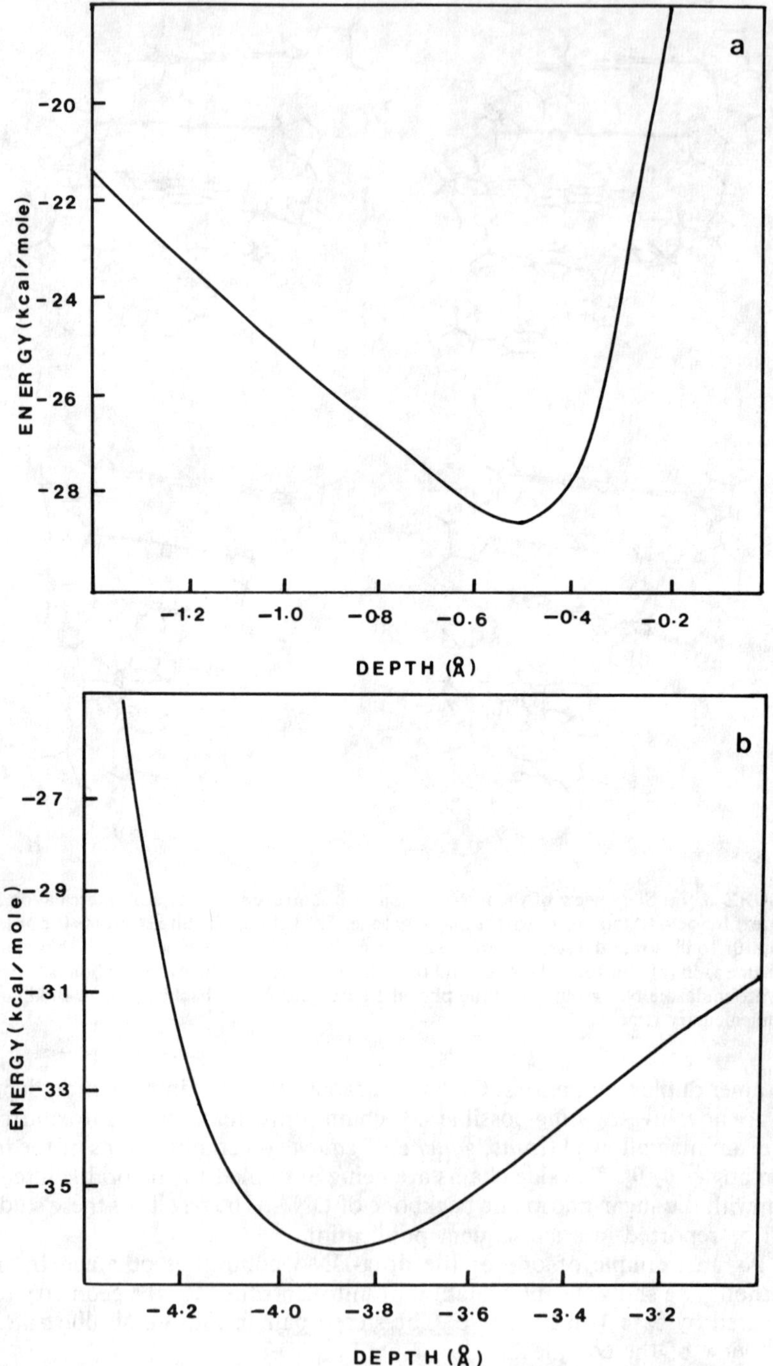
a
-20
-22
-24
-26
-28
ENERGY (kcal/mole)
-1.2
-1.0
-0.8
-0.6
-0.4
-0.2
DEPTH (Å)
b
-27
-29
-31
-33
-35
ENERGY (kcal/mole)
-4.2
-4.0
-3.8
-3.6
-3.4
-3.2
DEPTH (Å)

DNA-Bisantrene Intercalation

Preliminary studies using the graphics system revealed that bisantrene intercalation is much more favorable from the major groove side of the tetramer duplexes. Interactions of the side chains of bisantrene with the phosphate backbone severely limit the intercalation depth achievable from the minor groove side. This observation is supported by energy vs. depth calculations similar to those carried out for mitoxantrone above. Major groove intercalation is favored over minor groove by at least 10 kcal/mole in the cases investigated (TABLE 3).

In investigating major groove intercalation, we find that the angle θ_1 (see bisantrene) is restricted around $\pm$ ($90° \pm 40°$). The angle θ_2 is always *trans*. The angles θ_3 and θ_4 have considerable freedom, which leads to different orientations of the dihydroimidazole ring. The dihydroimidazole N(3)-H(7) groups can participate in hydrogen bonds with the main chain phosphate oxygens, thereby cross-linking the complementary strands of DNA. However, in such an intercalation complex, hydrogen bonds to a neighboring DNA helix are not possible, a result that is in agreement with the results from electron microscopy.[23] A stereo pair illustrating intercalation of bisantrene into the site geometry represented by case 1, above, is shown in FIGURE 8.

DESIGN OF NEW ANTICANCER DRUGS

Using the results summarized above, we have begun to investigate the design of new anticancer drugs. Given our experience with mitoxantrone and bisantrene and families of related compounds, we are focusing our design efforts on modifications to such compounds that might promote activity. The following strategies are being used to guide our efforts: (1) modifications to the planar ring systems to promote both major and minor groove binding; (2) modifications to the side chain functionalities to increase the chances for both intra- and interhelix cross-linking; (3) considerations of synthetic feasibility.

CONCLUSIONS

We have made significant progress toward understanding the geometries and energetics of the interactions of mitoxantrone and bisantrene with DNA tetramer duplexes. We have not, however, in the scope of this presentation, been able to illustrate all of our results. Nor have we been able to explore exhaustively potential intercalation geometries and drug side chain flexibilities. We are currently investigating the details of the energetics of side chain

---

FIGURE 9. (a) Conformational energy profile, in kcal/mole, as a function of depth of intercalation of mitoxantrone into a typical site (see text for details). **(b)** Same as **(a)** for intercalation of bisantrene.

TABLE 3. Minimum Intermolecular Energy and Corresponding Intercalation Depth for Minor and Major Groove Intercalation of Mitoxantrone and Bisantrene in Sites Referenced as Cases 1–5

| | Minor groove | | Major groove | |
| | Depth (angstroms) | Energy (kcal/mole) | Depth (angstroms) | Energy (kcal/mole) |
Case				
		Mitoxantrone		
1	−0.5	−28.4	5.0	−25.6
2	−1.0	−24.8	5.0	−27.6
3	0.0	−26.1	7.0	−16.5
4	0.0	−28.1	6.0	−21.0
5	−1.0	−25.1	6.0	−25.6
		Bisantrene		
1	−1.0	−22.2	6.0	−26.5
2	−1.0	−23.6	3.9	−36.2
3	−0.9	−22.3	6.2	−28.7
4	−0.4	−19.0	4.0	−27.0
5	−2.0	−19.9	4.8	−32.3

interactions with the phosphate backbones of representative intercalation geometries under conditions that will allow the system as a whole to relax under reasonable constraints. These results, including a more comprehensive presentation of twist-shift maps and alternative intercalation geometries, will be the subject of a future publication.

Our approach to the study of drug-DNA interactions has a number of limitations. Although we can locate energetically favorable intercalation geometries, we present no mechanistic explanation of how such geometries might be achieved in biological systems. Without such knowledge we cannot hope to distinguish the actual accessibility of such geometries, a classical quandary of kinetics vs. thermodynamics. Our methods cannot consider variations of all possible degrees of freedom in such complex systems because of limitations of computer time. Finally, although there is ample evidence that mitoxantrone and bisantrene are intercalators, there is no direct evidence that their efficacy as anticancer drugs results from their ability to interact with DNA in this way.

Despite these limitations, our modeling studies have yielded results that are in agreement with other observations. Both mitoxantrone and bisantrene are found to interact favorably with DNA duplexes, with a high degree of complementarity of steric and electrostatic surfaces and with formation of energetically favorable hydrogen bonds. We can now rationalize the ability of mitoxantrone, and the inability of bisantrene, to agglomerate DNA helices. Although we cannot at this stage explain in detail the observed *in vivo* structure/activity relationships of mitoxantrone, bisantrene, and related compounds, some of the gross differences in activity are apparent from our models. Certainly, our models are of sufficient utility to begin consideration of design of novel structures that would possess a high likelihood for activity.

SUMMARY

Flexibility of intercalation site geometries within a B-DNA helix was investigated in the twist-shift plane using energy minimization methods. The parameters optimized included sugar conformation, the glycosidic angles and phosphodiester torsion angles. Our calculations show several regions of energetically favorable intercalation geometries in the twist-shift plane. Modeling studies using interactive computer graphics and electrostatic potential surface compatibility provided initial hypotheses for the structures of the drug-DNA complexes. These hypotheses were supported and extended by energy minimizations of these complexes. Binding positions, conformational features and relative minimum binding energies of two anticancer drugs, mitoxantrone and bisantrene, were computed for intercalation complexes with DNA in the theoretically defined intercalation sites.

Mitoxantrone intercalates DNA from the minor groove and the side chain OH or NH groups are involved in hydrogen bonds with the main chain phosphate groups of DNA, thereby cross-linking the complementary strands. The hydroxyl groups of mitoxantrone can also participate in hydrogen bonding with phosphate oxygens of another chain, thereby cross-linking DNA helices. Bisantrene intercalates DNA favorably from the major groove and the NH group of the dihydroimidazole ring can participate in hydrogen bonding with the phosphate oxygens of the backbone. These models are consistent with the physicochemical and electron microscopic stuides of the interaction of mitoxantrone and bisantrene with DNA.

Our results are now being used to guide the design of novel anticancer drugs that should interact with DNA in a manner similar to that proposed for our representative drugs.

REFERENCES

1. ARCAMONE, F. 1983. *In* Structure-Activity Relationships of Anti-Tumor Agents. Developments in Pharmacology, vol. 3. D. N. Reinhoudt, T. A. Connors, H. M. Pinedo & K. W. van de Poll, Eds.: 111–133. Martinus Nijhoff. Boston.
2. HENRY, D. W. 1982. *In* New Approaches to the Design of Antineoplastic Agents. T. J. Bardos & T. I. Kalman, Eds.: 5–38. Elsevier. New York.
3. LERMAN, L. S. 1961. J. Mol. Biol. **3**: 18–30.
4. KREY, A. K. 1980. *In* Progress in Molecular and Subcellular Biology, vol. 7. F. E. Hahn, Ed.: 43–87. Springer-Verlag.
5. NEIDLE, S. & H. M. BERMAN. 1982. *In* Molecular Structure and Biological Activity. J. F. Griffin & W. L. Duax, Eds.: 287–302. Elsevier. New York.
6. NEIDLE, S. 1979. Progr. Med. Chem. **16**: 151–221.
7. NEIDLE, S. & H. M. BERMAN. 1983. Prog. Biophys. Molec. Biol. **41**: 43–66.
8. QUIGLEY, G. J., A. H.-J. WANG, G. UGHETTO, G. VAN DER MAREL, J. H. VAN BOOM & A. RICH. 1980. Proc. Nat. Acad. Sci. U.S.A. **77**: 7204–7208.
9. DREW, H. R., R. M. WING, T. TAKANO, C. BROKA, S. TANAKA, K. ITAKURA & R. E. DICKERSON. 1981. Proc. Nat. Acad. Sci. U.S.A. **78**: 2179–2183.
10. ALDEN, C. J. & S. ARNOTT. 1975. Nucl. Acids Res. **2**: 1701–1717.
11. ALDEN, C. J. & S. ARNOTT. 1977. Nucl. Acids Res. **11**: 3855–3861.
12. MILLER, K. J. 1979. Biopolymers **18**: 959–980.
13. MILLER, K. J. & J. F. PYCIOR. 1979. Biopolymers **18**: 2683–2719.

14. MAX, N. L. 1981. Comp. Chem. **5:** 19–27.
15. YATHINDRA, N., S. JAYARAMAN & R. MALATHI. 1983. *In* Conformation in Biology: Festschrift for G. N. Ramachandran. R. Srinivasan & R. H. Sarma, Eds.: 299–312. Adenine Press. New York.
16. GUPTA, G., M. H. SARMA & R. H. SARMA. 1983. Collected Abstracts: Conversation in Biomolecular Stereodynamics III. State University of New York at Albany, June 7–10. R. H. Sarma, Ed.: 27. Adenine Press. New York.
17. SRINIVASAN, A. R. & W. K. OLSON. Personal communication.
18. TAYLOR, E. R. & W. K. OLSON. Personal communication.
19. IUPAC-IUB Nomenclature. 1983. Eur. J. Biochem. **131:** 9–15.
20. MURDOCK, K. C., R. G. CHILD, P. F. FABIO, R. B. ANGIER, R. E. WALLACE, F. E. DURR & R. V. CITARELLA. 1979. J. Med. Chem. **22:** 1024–1032.
21. MURDOCK, K. C., R. G. CHILD, Y-I. LIN, J. D. WARREN, P. F. FABIO, V. J. LEE, P. T. IZZO, S. A. LANG, JR., R. B. ANGIER, R. V. CITARELLA, R. E. WALLACE & F. E. DURR. 1982. J. Med. Chem. **25:** 505–518.
22. FOYE, W. O., O. VAJRAGUPTA & S. K. SENGUPTA. 1982. J. Pharm. Sci. **71:** 253–257.
23. LOWN, J. W., C. C. HANSTOCK, R. D. BRADLEY & D. G. SCRABA. 1984. Mol. Pharmacol. **25:** 178–184.
24. KAPUSCINSKI, J., Z. DARZYNKIEWICZ, F. TRAGANOS & M. R. MELAMED. 1981. Biochem. Pharmacol. **30:** 231–240.
25. WALDES, H. & M. S. CENTER. 1982. Biochem. Pharmacol. **31:** 1057–1061.
26. OLSON, W. K. & P. J. FLORY. 1972. Biopolymers **11:** 25–56.
27. McGUIRE, R. F., F. A. MOMANY & H. A. SCHERAGA. 1972. J. Phys. Chem. **76:** 375–393.
28. LEVITT, M. 1983. Cold Spring Harbor Symposia on Quantitative Biology. **XLVII:** 251–262.
29. SRINIVASAN, A. R., N. YATHINDRA, V. S. R. RAO & S. PRAKASH. 1980. Biopolymers **19:** 165–171.
30. BALAJI, V. N. & W. K. OLSON. Unpublished data.
31. BINGHAM, R. C., M. J. S. DEWAR & D. H. LO. 1975. J. Am. Chem. Soc. **97:** 1285–1293.
32. MANNING, G. S. 1979. Accts. Chem. Res. **12:** 443–449.
33. ARNOTT, S. & D. W. L. HUKINS. 1973. J. Mol. Biol. **81:** 93–105.
34. ARNOTT, S. 1970. *In* Progress in Biophysical and Molecular Biology, vol. 21. J. A. V. Butler & D. Noble, Eds: 265–319. Pergamon. New York.
35. ARNOTT, S. & D. W. L. HUKINS. 1972. J. Biochem. **130:** 453–465.
36. ARNOTT, S. & D. W. L. HUKINS. 1972. Biochem. Biophys. Res. Commun. **47:** 1504–1509.
37. SASISEKHARAN, V. 1973. *In* Conformation of Biological Molecules and Polymers. E. D. Bergmann & B. Pullman, Eds.: 247–260. The Israel Academy of Sciences and Humanities. Jerusalem.
38. PATTABIRAMAN, N., S. N. RAO & V. SASISEKHARAN. 1980. Nature **284:** 187–189.
39. FRATINI, A. V., M. L. KOPKA, H. R. DREW & R. E. DICKERSON. 1982. J. Biol. Chem. **257:** 14686–14707.
40. ALLEN, F. H., S. BELLARD, M. D. BRICE, B. A. CARTWRIGHT, A. DOUBLEDAY, H. HIGGS, T. HUMMELINK, B. G. HUMMELINK-PETERS, O. KENNARD, W. D. S. MOTHERWELL, J. R. RODGERS & D. G. WATSON. 1979. Acta Crystallogr. **B35:** 2331–2339.
41. NIGAM, G. D. & B. DEPPISCH. 1980. Z. Kristallogr. **151:** 185–191.
42. HENRY, D. W. 1976. Cancer Chemotherapy. A. C. Sartorelli, Ed.: 15–57. American Chemical Society. Washington.

43. NEIDLE, S. 1978. *In* Topics in Antibiotic Chemistry, vol. 3. P. G. Sammes, Ed: 249–278. Ellis Horwood. Chichester.
44. PIGRAM, W. J., W. FULLER & L. D. HAMILTON. 1972. Nature New Biol. **235**: 17–19.
45. NEIDLE, S. 1977. Cancer Treat. Rep. **61**: 923–929.
46. HOPFINGER, A. J. & Y. NAKATA. 1981. *In* Intermolecular Forces. B. Pullman, Ed.: 431–444. Reidel. Dordrecht, the Netherlands.
47. O'DONNEL, T. J. & A. J. OLSON. 1981. Computer Graphics **15**: 133–142.
48. JONES, T. A. 1978. J. Appl. Crystallogr. **11**: 268–272.
49. WEINER, P. K., R. LANGRIDGE, J. M. BLANEY, R. SCHAEFER & P. A. KOLLMAN. 1982. Proc. Nat. Acad. Sci. U.S.A. **79**: 3754–3758.
50. CONNOLLY, M. L. 1983. Science **221**: 709–713.

Structure-Activity Studies:
A Three-Dimensional Probe
of Receptor Specificity

GARLAND R. MARSHALL

Department of Physiology and Biophysics
Washington University School of Medicine
St. Louis, Missouri 63110

INTRODUCTION

In the wake of the increased number of physiologically important proteins whose three-dimensional structures are being determined and the sophisticated computer graphics used in their study, one can lose sight of the common problem that faces most investigators interested in the study of receptors. One must infer a hypothetical picture based on the properties that can be measured, often requiring the ultimate in sensitivity in methodology because of the minute quantities of receptor under study.

By varying the structure of the ligand, one can probe the receptor and attempt to separate the variables that lead to recognition and activation of the particular receptor under study. This approach requires numerous assumptions, some of which are often tacit, and can lead to erroneous concepts if the particular assumptions are invalid. Only by consistency with the set of observations and by prediction of new and testable results can any model derive credibility. Continued refinement of the hypothetical receptor model in the face of new data is obligatory; one must avoid emotional attachment to any particular model, as most are underdetermined. In spite of these caveats, several useful applications of the use of structure-activity studies to probe receptor specificity have appeared. Some of these, as well as studies from our own group, will be used to illustrate the general approach.

PHARMACOPHORE HYPOTHESIS

One oversimplification which is, nevertheless, a good starting point is the traditional concept of pharmacophore, normally credited to Erlich at the turn of the century. In more modern terms, the pharmacophore is the critical part of the key which interacts with the tumblers necessary to unlock the appropriate events leading to receptor activation. In other words, it is a unique three-dimensional pattern of electron density in the ligand which activates the receptor. Other than Occam's razor, there are no *a priori* reasons why one should conceive of a unique pattern. In fact, examination of crystal structures of enzyme-substrate/inhibitor complexes such as those seen with dihydrofolate reductase indicates considerable variation in the location and nature of interacting groups, although the three-dimensional arrangements of groups within the active site with which the ligands interact do appear to be relatively consistent. If a consistent and predictive model can be derived without the introduction of multiple binding modes or induced fit, it has obviously fewer assumptions and is preferred until the data force an increase in the complexity of the model.

ACTIVE ANALOG APPROACH

One can assume that a single pharmacophore exists for a set of compounds active at a given pharmacological receptor, and use a logical approach to help determine if such a simplified concept is consistent with the observed data. For example, inhibitors of angiotensin-converting enzyme all seem to have a *C*-terminal carboxylate, an amide bond, and a group capable of binding to zinc. One could examine the possible three-dimensional arrangements of these three groups available to each analog to see if a common pattern were accessible. One approach would be to link each of the groups together across the series of analogs and simultaneously minimize the resulting

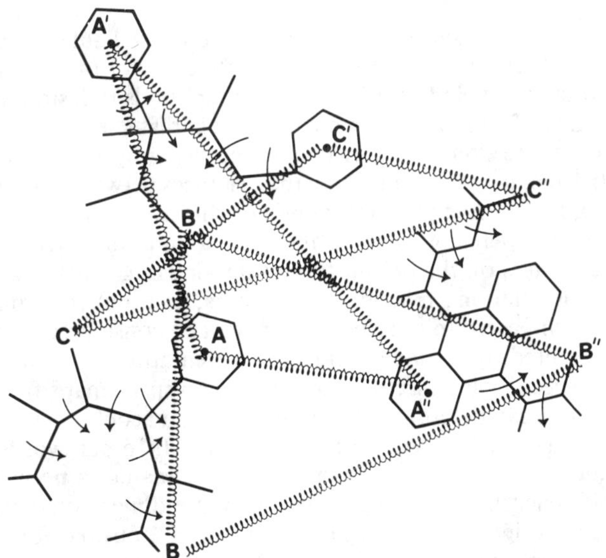

FIGURE 1. Minimization of "supermolecule" approach. Groups with assumed common modes of interaction are linked by "spring" and molecules minimized simultaneously subject to these constraints while otherwise invisible to the other molecules.

supermolecule with each molecule invisible to the others (Fig. 1). By appropriate weighting of the spring constants, the molecules can be made to assume similar three-dimensional arrangements of the three groups.[1] Unfortunately, there are two major deficiencies in such an approach. First, the results are dependent on the starting positions, as only the nearest local minima in the total energy function is found. Second, no information regarding the global minima nor the uniqueness of the solution found is available. It is obviously important to have an estimate of the number of possible pharmacophore hypotheses consistent with the set of assumptions as well as the existence of a possible solution.

Because the constraints on a molecule to induce the appropriate biological response are more stringent than simply high affinity for the active site, which could lead to inhibition only, one has a higher probability of success with the pharmacophore assumption with agonists than with antagonists. One approach to the question of whether agonists and antagonists bind to the same site in the same state would be analysis of antagonist and agonist data separately and comparison of the resulting models to see if a consistent view arises. Klunk *et al.*[2] have recently made such an independent analysis for picrotoxinin and γ-butyrolactone convulsants. Their data is consistent with a common binding mode at a common receptor with the lactone ring in both sets of compounds interacting in a similar fashion (Fig. 2).

ORIENTATION SPACE

In order to search efficiently for common patterns available to a set of drugs interacting with the same receptor, a methodology[3] independent of the molecular framework and rotation of the compounds is desirable. By transforming each conformation (which could be expressed either in terms of internal degrees of freedom, *i.e.*, torsional rotations, or by absolute coordinates) into a relative distance space in which the distances between pharmacophoric groups becomes the metric, compounds of different congeneric series can be compared independent of their absolute orientations. This relative distance space focuses on the *orientation* of important groups relative to each other, *i.e.*, potential pharmacophoric patterns, and thus the name, orientation space. A three-group pharmacophore can be considered a triangle that can be represented by a single point in an orientation space with a dimensionality of three. By intersecting the orientation space maps for a set of active analogs, one can determine if a common pharmacophoric pattern or patterns can be assumed by each of the analogs. In other words, for each sterically allowed conformation available to an analog, a point in orientation space is generated. This represents the three-dimensional arrangement of the groups designated by the chemist to be responsible for activity for that particular conformer. The set of such points represents all the possible arrangements available to the analog. Since our hypothesis of a pharmacophore implies a common pattern among the set of active analogs, the corresponding point in orientation space must be occupied in each orientation space

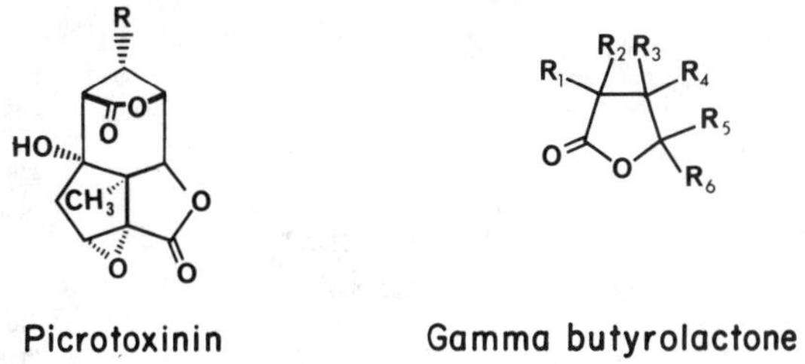

Picrotoxinin   Gamma butyrolactone

FIGURE 2. Convulsant and anticonvulsant molecules that share a lactone ring assumed to be common binding element.

map generated for each analog. The common set of points represents the set of pharmacophore patterns consistent with the assignments of pharmacophoric groups and the set of analogs examined.

RECEPTOR MAPPING

Besides the ability to present an appropriate message, the pharmacophore, other criteria are necessary for activity. Sufficient affinity must be present in order to distinguish the activity over background noise attributable to nonspecific interactions, and this affinity can be severely compromised by a negative steric interaction with the receptor. In other words, the correct pattern may be present on the key, but the key blank may not fit the lock. Comparison of the set of key blanks that work will outline some of the space avail-

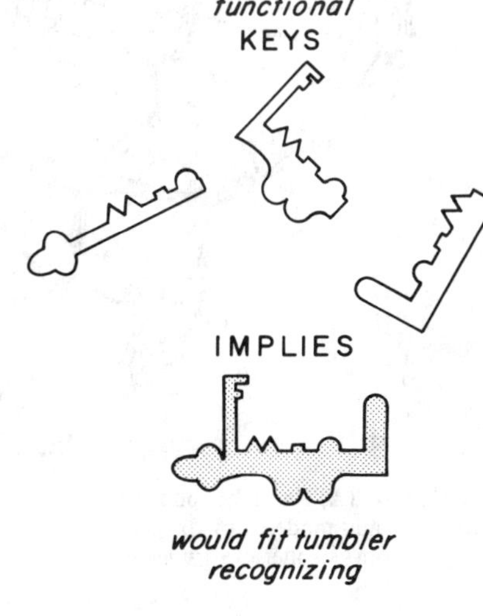

FIGURE 3. Assumption underlying receptor mapping illustrated with key and tumbler analogy.

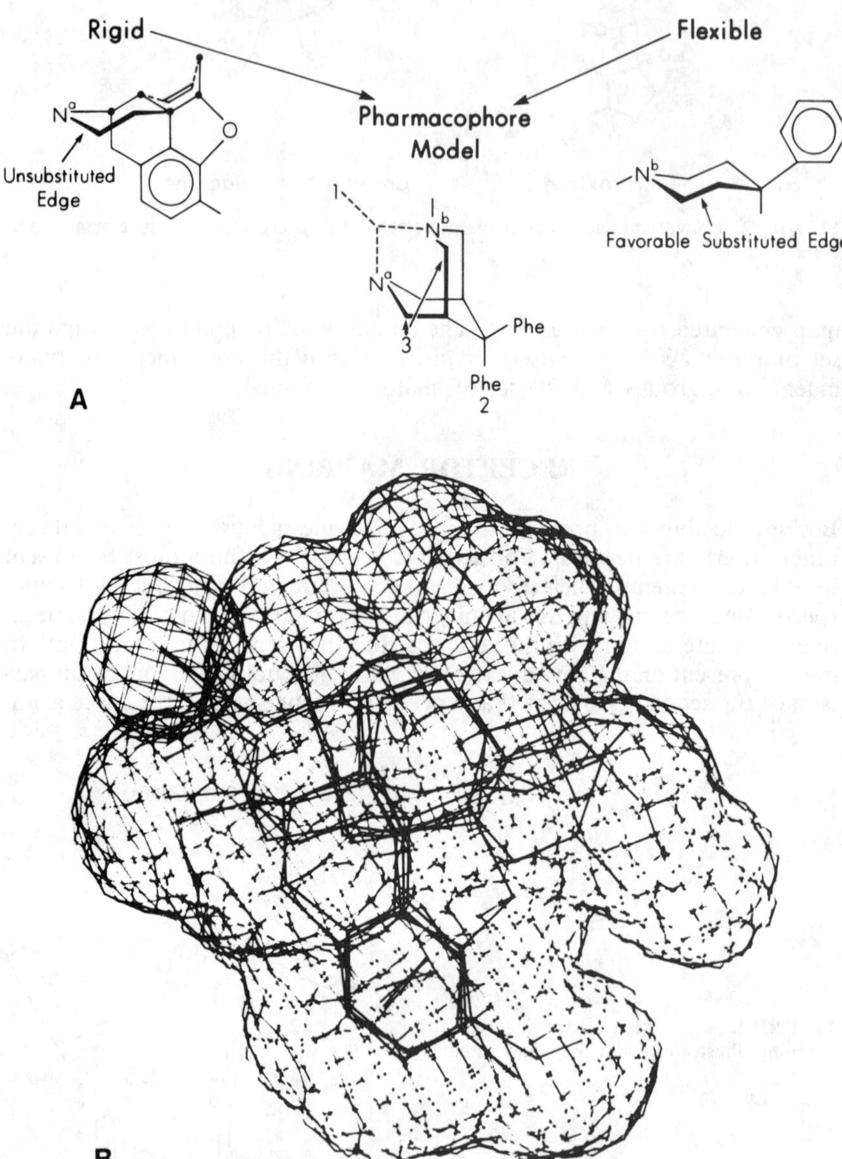

FIGURE 4. (A) Pharmacophore model for opioid μ-receptor with nitrogen binding site indicated by *1* and modeled by oxygen atom. (B) Excluded volume map based on pharmacophore; oxygen atom is sphere in *upper left*.

able for successful key blanks when aligned with the common pattern (FIG. 3). The assumption is made, therefore, that the common pharmacophoric pattern represents a common binding mode and can be used to align the set of analogs in a common frame of reference. The union of the volume of the parts of each analog which are fixed during presentation of the pharmacophore defines a minimal volume available at the receptor.

Based on the above arguments, one reason for inactivity could be the competition between the receptor and the analog for common space when the analog has assumed the appropriate conformation to present the pharmacophore. This becomes another basis for a check of consistency as well as an indication of the receptor orientation relative to the pharmacophore. If an inactive analog requires no novel volume while presenting the pharmacophore besides that defined by the union of actives, then something is wrong somewhere. The assumption of a common mode of action, the trial pharmacophore hypothesis, assignment of correspondence between functional groups, and/or measurement of activity all become suspect. In general, an alternative pharmacophore hypothesis is explored for consistency. If there is good pharmacological data suggesting a common site of action, *i.e.*, competitive binding assays, and either a null set occurs for the orientation map analysis or an inconsistency cannot be avoided in receptor mapping, then a more complicated interpretation, *e.g.*, multiple binding modes, must be explored.

Several methods to introduce quantitation into such receptor maps are on the horizon. Crippen has developed the distance geometry approach[4] to give good quantitative correlation between observed affinity and a derived site model. Although this is clearly an excellent method which deserves further development and wider application, it does suffer from its dependence on minimization, with both the local minima problem and questions of uniqueness. When cycled in a prediction, synthesis, and testing refinement, this procedure should prove both powerful and practical. Wise and Cramer[5] have developed an approach, DYLOMMS (the dynamic lattice method), which has the potential to accomodate both conformeric and tautomeric averaging. The simple assumption of additivity in drug-receptor interactions should allow one to partition observed binding energy based on structural modification. Besides the difficulty of separation of variables, *i.e.*, a modification changes steric, lipophilic, and electrostatic factors, measured affinities are related to free energy, not enthalpy, and entropic contributions including solvation cannot be dismissed.

BEYOND THE PHARMACOPHORE

As has been discussed previously, the strict pharmacophore hypothesis is clearly an oversimplification of what our experience would suggest to be the general case. Receptor sites are multipotential with respect to modes of binding of a single conformer of a drug. The next level of assumption is to assume that the receptor site is relatively static, and explore how one can

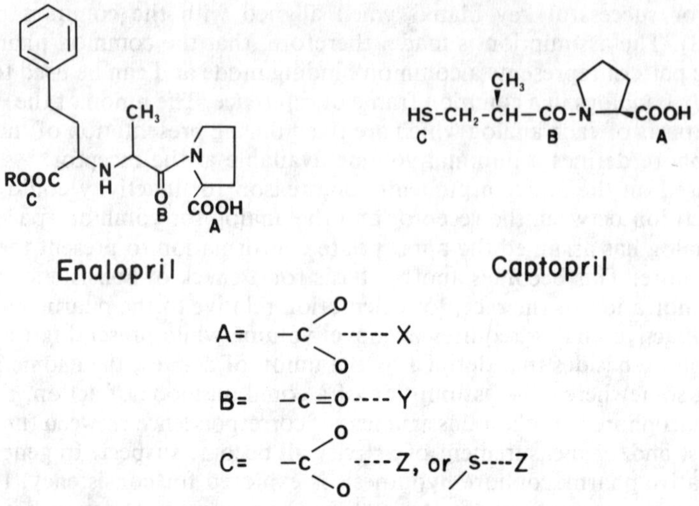

FIGURE 5. Proposed binding site extensions for two angiotensin-converting enzyme inhibitors.

derive site models consistent with the available data. The approach developed by Crippen allows user specification of site points[4] which could be computer-assisted. One can extend the methodology of the active analog approach by systematically searching for coincidence of sites capable of interacting with functional groups rather than for overlapping of those functional groups. To give an example, detailed consideration of the opioid μ-receptor by Portoghese[6] has led to a proposal of multiple binding modes for substituted piperadine opioids. Humblet and Marshall[7] were led to a similar conclusion by extending the lone pair on the nitrogen to a distance capable of hydrogen bonding to an oxygen atom and searching for overlap of the postulated receptor site oxygen. This plus an additional phenyl binding site which was clearly indicated by structure-activity data forms the current basis of their model of the μ-receptor (FIG. 4). Such an extension is clearly justified in the case of angiotensin-converting enzyme inhibitors by extending the sulfhydry function of captopril and the carboxylate functionality of enalapril to include the zinc atom (FIG. 5). Such an analysis of ACE inhibitors is underway (Schneider and Marshall, unpublished).

Why not include such elaborations in the initial analysis? The number of assumptions is clearly greater. The number of degrees of freedom for systematic search have to be increased, which can be computationally crippling. But more importantly, the accumulated structure-activity data often are not sufficient to rule out a much more simple hypothesis. What one is attempting is to use analogs to probe the receptor site in an effort to deduce as much as possible about its properties. Using this technique (similar to the parlor game of Twenty Questions), one can clearly derive a great deal of information by careful selection of questions and pruning of the combinatorial tree of possibilities depending on the answers. There are many systems, such as the opioid or dopamine receptor, where the medicinal chemists have synthesized and had the activity measured on thousands of compounds. Clearly, we are novices at this version of Twenty Questions, but glimmers of a successful strategy are available.

REFERENCES

1. Duchamp, D. J. 1977. Molecular mechanics and crystal structure analysis in drug design. *In* Computer-Assisted Drug Design. E. C. Olson & R. E. Christoffersen, Eds. ACS Symposium Series **112**: 79–102. American Chemical Society. Washington, D.C.
2. Klunk, W. E., B. L. Kalaman, J. A. Ferrendelli & D. F. Covey. 1983. Mol. Pharmacol. **23**: 511–518.
3. Marshall, G. R., C. D. Barry, H. E. Bosshard, R. A. Dammkoehler & D. A. Dunn. 1977. The conformational parameter in drug design: The active analog approach. *In* Computer-Assisted Drug Design. E. C. Olson & R. E. Christoffersen, Eds. ACS Symposium Series **112**: 205–226. American Chemical Society. Washington, D.C.
4. Crippen, G. M. 1981. Distance geometry and conformational calculations. *In* Chemometrics Research Studies Series, vol. 1. D. Bawden, Ed. Wiley. Chichester, England.
5. Wise, M., R. D. Cramer, D. Smith & I. Exman. 1983. *In* Quantitative Approaches to Drug Design. J. C. Dearden, Ed. Elsevier. Amsterdam, the Netherlands.
6. Fries, D. S. & P. S. Portoghese. 1976. J. Med. Chem. **19**: 1155.
7. Humblet, C. & G. R. Marshall. 1981. Drug Development Res. **1**: 409.

PART V. IMPACT OF COMPUTER HARDWARE & GRAPHICS TECHNOLOGY: FUTURE PERSPECTIVES

Calculations of the Three-Dimensional Structures of Proteins[a]

HAROLD A. SCHERAGA
Baker Laboratory of Chemistry
Cornell University
Ithaca, New York 14853

INTRODUCTION

We are using a combined theoretical and experimental approach to gain an understanding of the interactions that dictate how polypeptide chains fold into the conformations of native proteins and then (*e.g.*, as enzymes) interact with substrates. The theoretical and experimental studies complement each other, each providing information needed and used by the other. This paper is concerned primarily with our theoretical (computational) approach, but some experimental results will be cited where relevant.

Computational algorithms will not be discussed here because they have been reported and summarized in a number of papers (*e.g.*, in ref. 1). These algorithms include procedures for treating entropy effects [2–6] and hydration effects.[7–9] The agreement between computational and experimental results for both small and large polypeptides,[10] to be discussed herein, serves to indicate the reasonableness of the computational procedures (including minimization algorithms) and of the parameters of the empirical potential energy functions.

The main problem on which to focus is that which arises from the presence of many local minima in the complex, multidimensional energy surface, *i.e.*, the multiple-minima problem. If the number of variables is small, this problem can be surmounted by a build-up procedure that has some features in common with the real-space renormalization group technique.[11] As the number of variables increases, this procedure is augmented by several short-, medium-, and long-range approximations to reach the potential well in which the global minimum lies.[11] These procedures enable the conformational space

FIGURE 1. Illustration of three Tyr . . . Asp interactions in bovine pancreatic ribonuclease A, deduced from various chemical and physical chemical experiments.[12] The drawing was prepared from the X-ray coordinates of Wlodawer *et al.*[13]

[a] This work was supported by research grants from the National Institutes of Health (no. GM–14312), the National Science Foundation (no. PCM79–20279), and the National Foundation for Cancer Research.

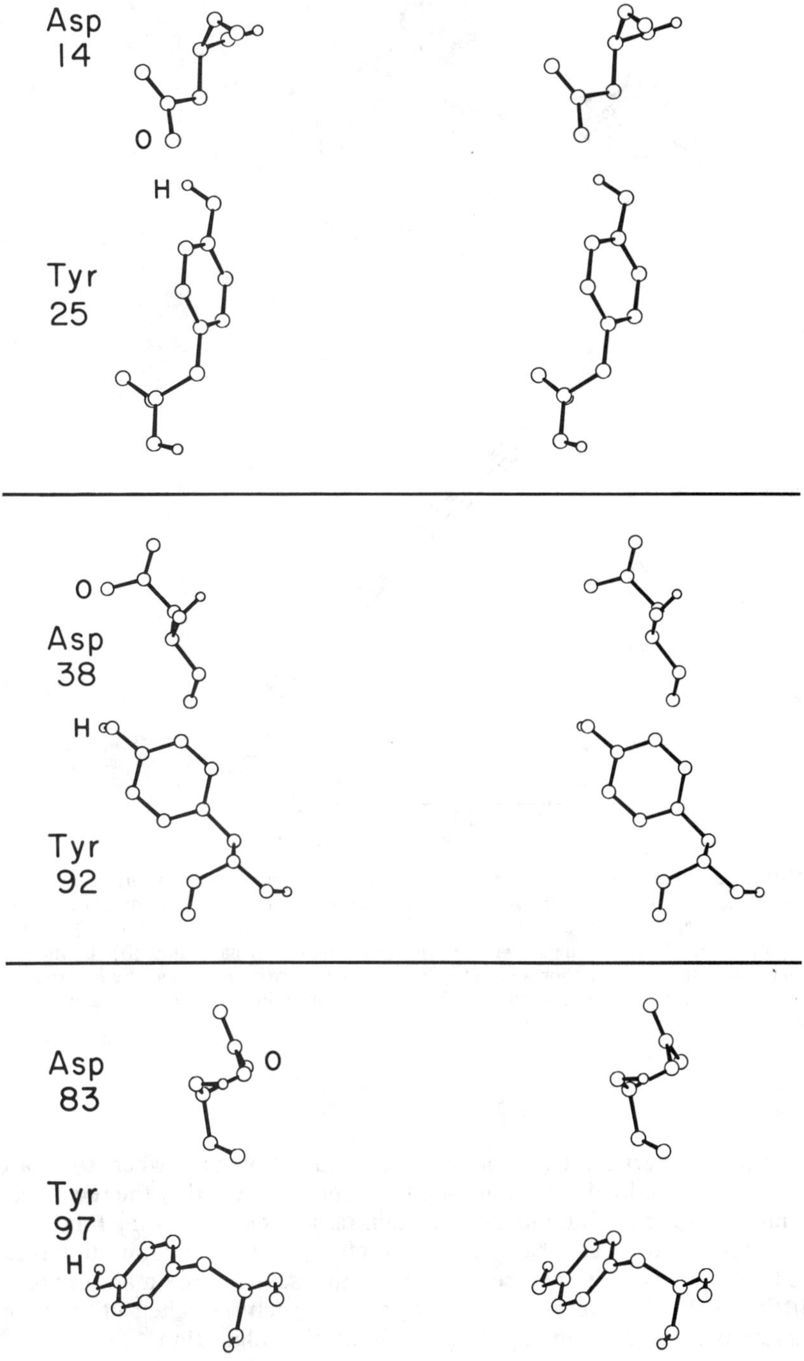

Asp
14
O
H
Tyr
25
O
Asp
38
H
Tyr
92
Asp
83
O
Tyr
97
H

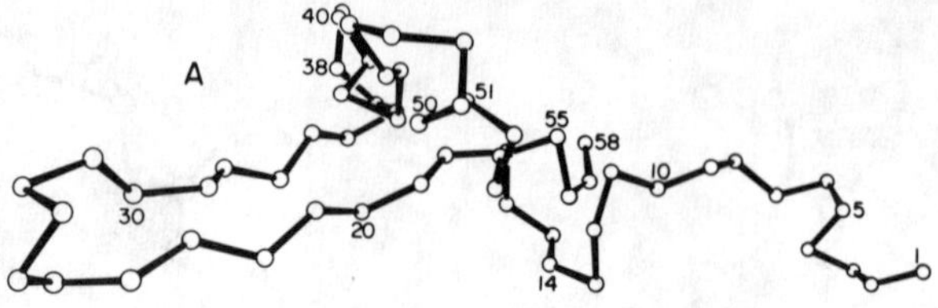

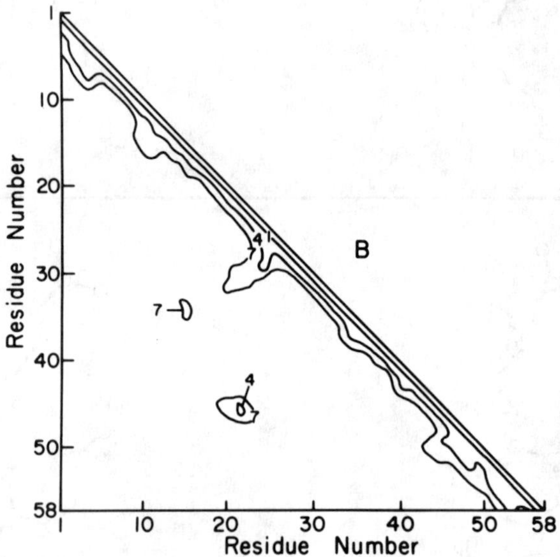

FIGURE 2. Structural comparisons for BPTI.[20] **(A)** Spatial structure with correct *short-range* order, generated by assigning to each residue the *average* value of the backbone dihedral angles ϕ and ψ in its appropriate conformational state (obtained from the X-ray structure). Only the positions of the C^α atoms are shown as circles connected by virtual bonds. **(B)** Distance contours in Å of the structure shown in **(A)**. **(C)** Spatial structure as determined by X-ray crystallography. **(D)** Distance contours in Å of the structure shown in **(C)**. The numbers on the contours should all be augmented by 2 Å.

to be searched efficiently, avoiding large volumes of space where the search would be unproductive and, at the same time, assuring that the region containing the global minimum is reached in reasonable computing time. Then the approximations are abandoned, and full-scale energy minimization is carried out. However, it must be emphasized that such approximations are not sufficient, by themselves, to predict protein structures. Their utility is primarily as constraints in an energy-minimization algorithm.

FIGURE 2 *(continued)*

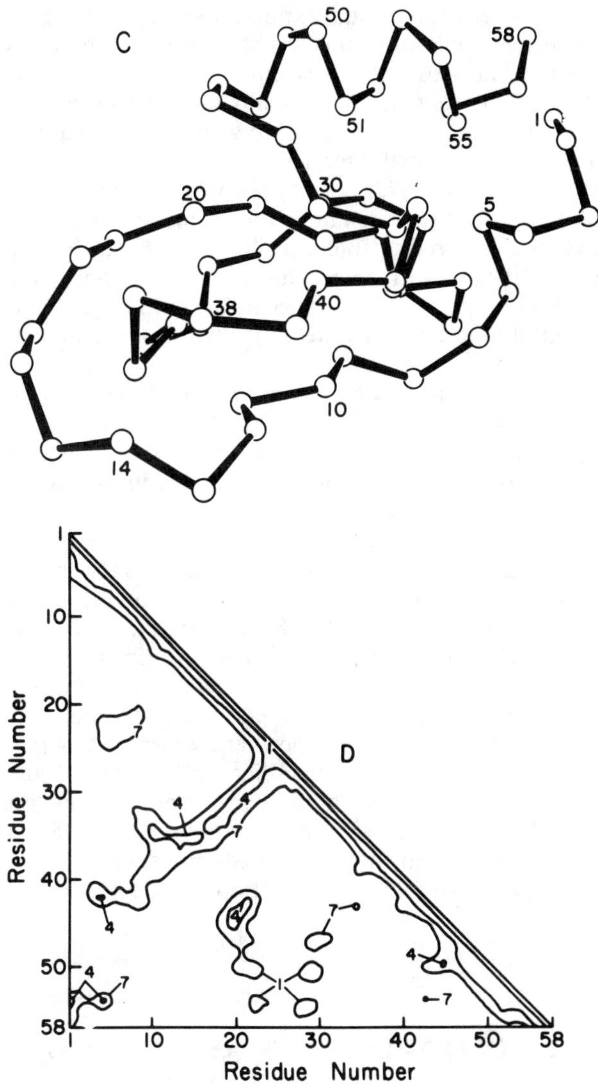

USE OF EXPERIMENTAL RESULTS AS CONSTRAINTS

Several years ago, we used various chemical and physical chemical procedures to identify three tyrosyl . . . aspartyl interactions in ribonuclease.[12] When the X-ray structure was subsequently determined,[13] these three interactions were found to be present, as shown in FIGURE 1. Considering that there are 19,800 ways to pair three out of six tyrosyl groups with three out of eleven carboxyl groups, it is seen that the set of pairings of FIGURE 1 represents a unique result. Nowadays, nuclear Overhauser[14,15] and nonradiative energy transfer[16] measurements are being used to obtain similar information. With the knowledge of these three distance constraints, and the known location of the four disulfide bonds of ribonuclease, we began the development of theoretical procedures[17] to generate polypeptide chains and compute their conformational energies.

However, as recent analyses[18,19] have shown, an enormous number (of the order of hundreds) of such distance constraints (covering both small and large distances) are required to fold a protein as small as bovine pancreatic trypsin inhibitor (BPTI), with 58 residues, if only distance constraints are used. On the other hand, if such distance information is used as constraints in an optimization algorithm (see below), then many fewer distance constraints are required.

Other useful information, which can serve as constraints in an energy-minimization algorithm, is knowledge of the conformational states of each residue of a protein; *e.g.,* whether or not a residue is in an α-helical or some other state, based on short-range prediction procedures. Such approximations have limited accuracy, but nevertheless are helpful *if used together with energy minimization*. However, by themselves, they give a false picture of protein structure. Even if they were 100% accurate, short-range prediction algorithms would give an entirely different description of a protein than the actual one, as illustrated[20] for BPTI in FIGURE 2; *i.e.,* the conformations of FIGURES 2A and 2C have the same local structure (in the sense that corresponding residues are in the same local region of a ϕ, ψ conformational map), but they differ considerably in their long-range interactions.

The situation can be improved somewhat by introducing medium-range interactions. For example, the predictability of the helical and extended-structure regions is improved considerably by introducing medium-range interactions up to four residues from the one under consideration,[21] as illustrated in FIGURE 3. But, even short- and medium-range interactions are not sufficient by themselves to fold a protein. Long-range interactions must also be included.

USE OF EXPERIMENT AS A TEST OF COMPUTATIONAL RESULTS

Numerous experimental tests of our computational results have been carried out, and several examples are cited here for illustrative purposes. The systems studied range from simple models of α-helices and β-sheets to oligopeptides to globular proteins. In essentially all cases, the most probable state was attained by energy minimization, taking all interactions into account.

A predicted variant[22] of Pauling's α-helix (*i.e.,* α_I-helix), designated α_II (with NH $\cdots$ O $=$ C hydrogen bonds that are bent, in constrast to the near linear ones of the α_I helix) was found to exist to some extent in myoglobin

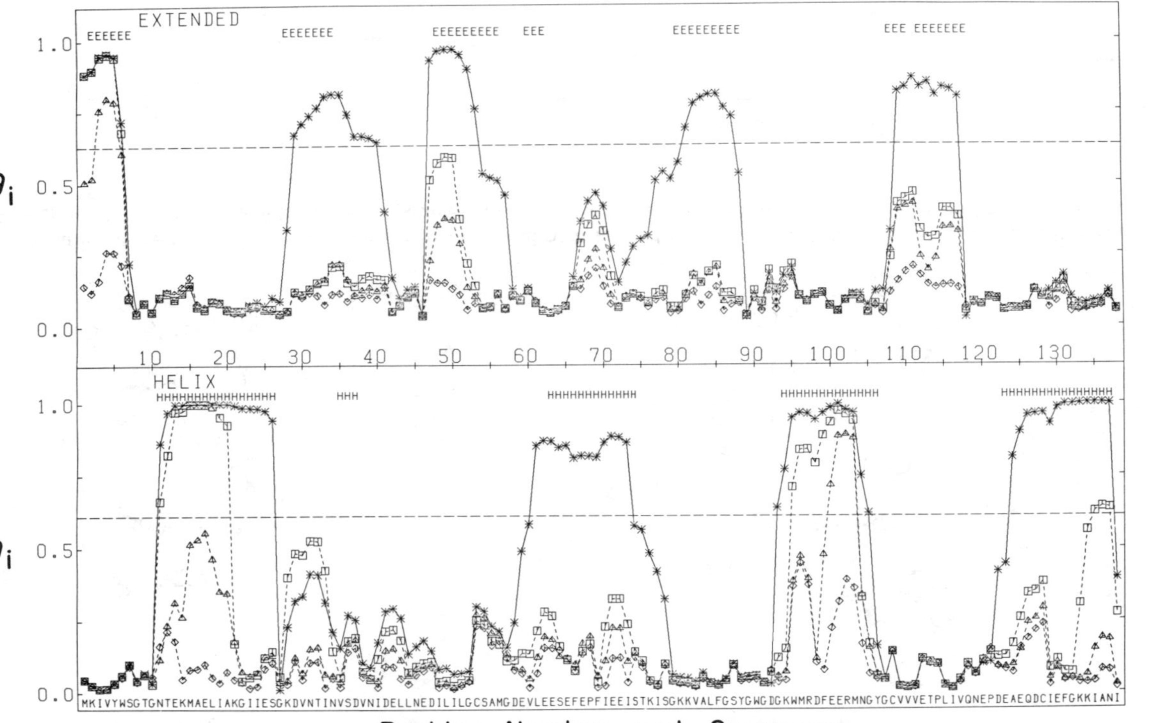

FIGURE 3. Calculated α-helix and extended-structure probability profiles for flavodoxin.[21] The symbols $\diamond$, $\triangle$, $\square$, and * represent profiles that take into account interactions up to 1, 2, 3 and 4 neighbors, respectively. The dashed lines are thresholds to predict the conformational state of each residue as either an α-helix (or extended structure) or a coil. The letters H and E indicate the residues that are in the α-helical and extended-structure states, respectively, in the native conformation. The amino acid sequence of flavodoxin is shown in the one-letter representation at the bottom. [Reprinted with permission from ref. 21.]

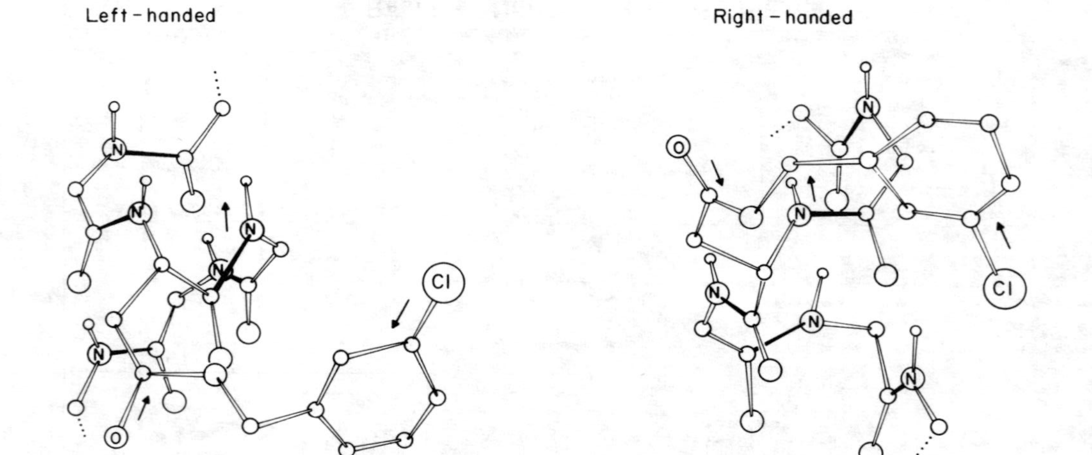

FIGURE 4. Orientation of the side chains of the lowest energy left- and right-handed α-helices of poly(m-Cl-benzyl-L-aspartate)[24]. The *arrows* represent the directions of the C-Cl, ester, and amide dipoles, respectively. [Reprinted with permission from ref. 24. Copyright 1970 American Chemical Society.]

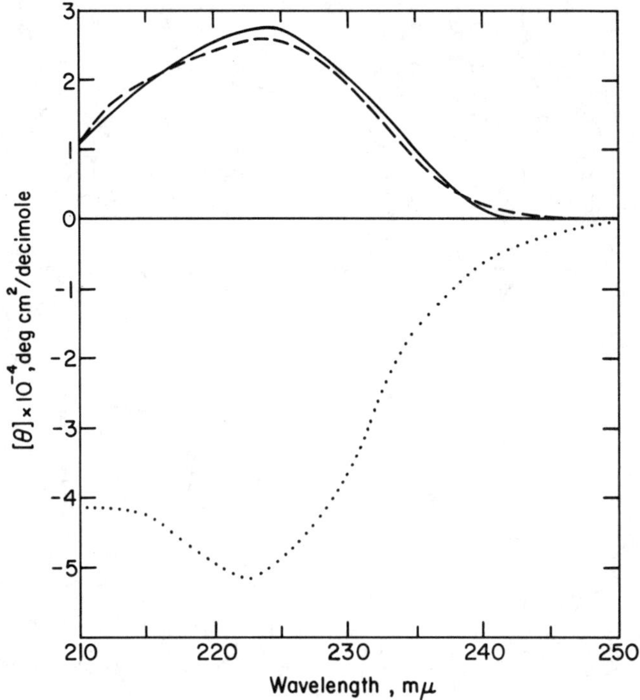

FIGURE 5. Circular dichroism spectra[25] of poly-β-benzyl-L-aspartates in dioxane at 25°C. The symbols (——), (– – –), and (· · ·) correspond to the o-, m-, and p-Cl-benzyl derivatives, respectively. [Reprinted with permission from ref. 25. Copyright 1970 American Chemical Society.]

and lysosyme.[23] The right- or left-handed twist of numerous α-helices was predicted and verified experimentally. Moreover, the interactions that determine the helix sense were identified. For example, as shown in FIGURE 4, side chain-backbone dipole-dipole interactions play a dominant role in leading to a left-handed helix in poly(m-Cl-benzyl-L-aspartate).[24] The predicted[24] left-handedness of the o- and m-derivatives, and the right-handedness of the p-derivative, were subsequently verified by circular dichroism measurements,[25] as indicated in FIGURE 5. Similarly, energy minimization has led to a preference for right-handed β-sheets,[26-30] as observed in globular proteins by Chothia,[31] and the interactions leading to this preference have been identified. FIGURE 6 illustrates the computed right-handed twist of poly-L-valine β-sheets.[27] Similar computations have been carried out for a variety of other homopolymers and sequential copolymers of amino acids.[30]

In considering a molecular theory of the helix-coil transition in homopolymers, it was necessary to compute the free energy of both the helix and coil to simulate the transition curves in a given solvent. FIGURE 7 shows how the computations[32] match the experimentally observed[33] increase in the Zimm-

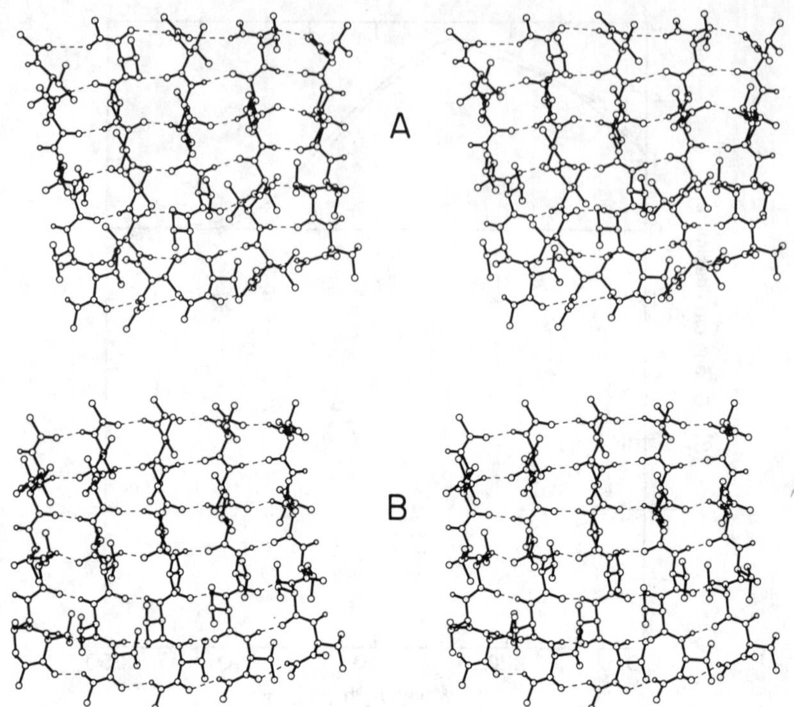

FIGURE 6. Stereo drawings of the minimum-energy β-sheets with five CH_3CO-$(L$-$Val)_6$-$NHCH_3$ chains.[27] (**A**) Antiparallel structure. (**B**) Parallel structure.

Bragg[34] helix growth parameter s with temperature for poly(L-valine) in water.

An analog of the Zimm-Bragg matrix treatment of the helix-coil transition has recently been developed for the β sheet-coil transition.[35] As illustrated in FIGURE 8, allowance is made for any number of residues per strand, of strands per sheet, and of sheets per chain. Edge effects (*i.e.*, the presence of residues in the outer strands or in the turns of a sheet), which lead to the analog of the Zimm-Bragg parameter σ, play an important role in determining the average size of a β-sheet. Such a treatment can extend the predictability of extended structures (FIGURE 3) to that of antiparallel β-sheets.

Computational results on fibrous proteins, such as collagen and synthetic polypeptide models thereof, have also been verified by experiment. For example, the predicted structure of the collagen-like triple-stranded helix of poly(Gly-Pro-Pro),[36] shown in FIGURE 9, has Cartesian coordinates that agree with those from a single-crystal X-ray diffraction study[37] of $(Pro$-Pro-$Gyl)_{10}$ within ± 0.3 Å. The computed results for other polytripeptides[38-40] are also in agreement with experiment.

Recently, we have applied the same methodology to $\alpha \ldots \alpha$, $\alpha \ldots \beta$, and $\beta \ldots \beta$ interactions in globular proteins in order to gain an understanding as to how these structural complexes arise. For example, FIGURE 10 shows

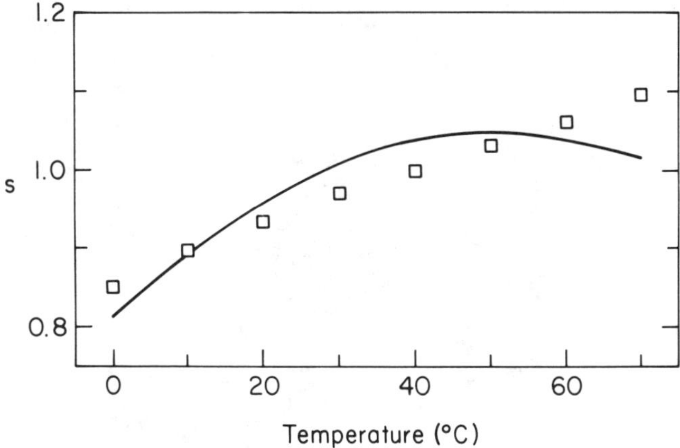

FIGURE 7. Comparison of s vs. T curves for poly(L-valine) in water. The line is a calculated one,[32] and the squares are the experimental results.[33] [Reprinted with permission from ref. 32. Copyright 1974 American Chemical Society.]

the lowest-energy packing state of two poly(L-alanine) α-helices.[41] Some of the low-energy packing arrangements, arrived at by energy minimization, could not be deduced from geometrical considerations (such as "knobs-in holes"[42]) alone. Similar methods have been applied to the assembly of triple-helical structures of collagen-like polytripeptides.[28,43]

In the intermediate-size range of oligopeptides, gramicidin S serves as an example. FIGURE 11A shows the computed structure[44,45] of this cyclic

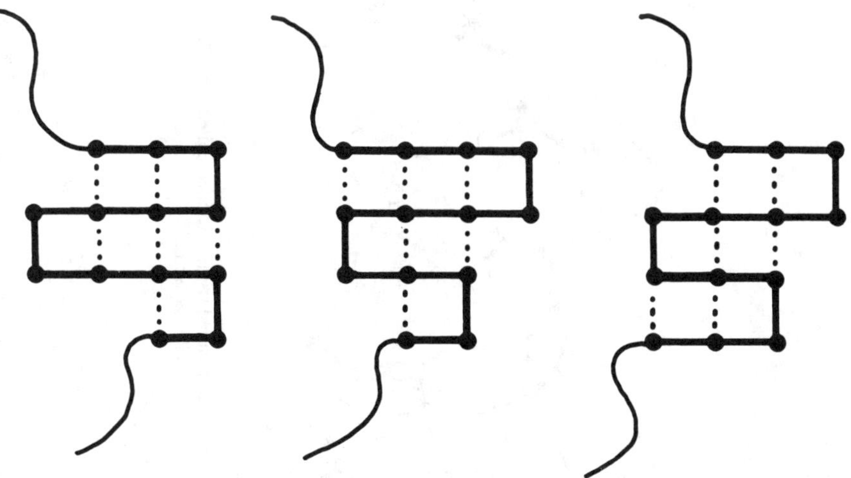

FIGURE 8. Illustration of various antiparallel β-sheets.[35]

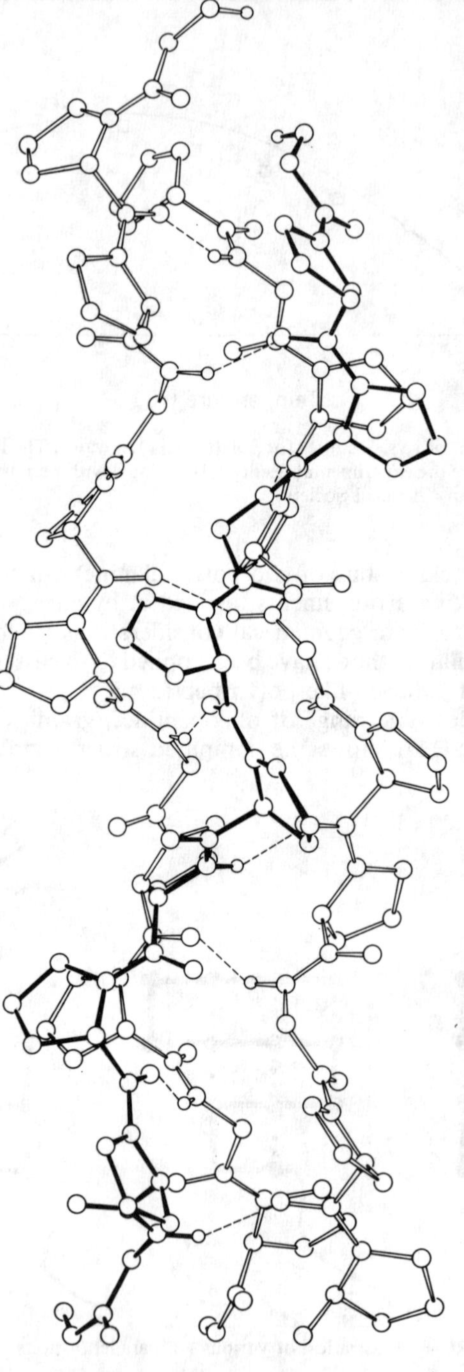

FIGURE 9. Computed triple-stranded coiled-coil complex of poly(Gly-Pro-Pro) of lowest energy.[36] [Reprinted with permission from ref. 36. Copyright 1976 John Wiley & Sons.]

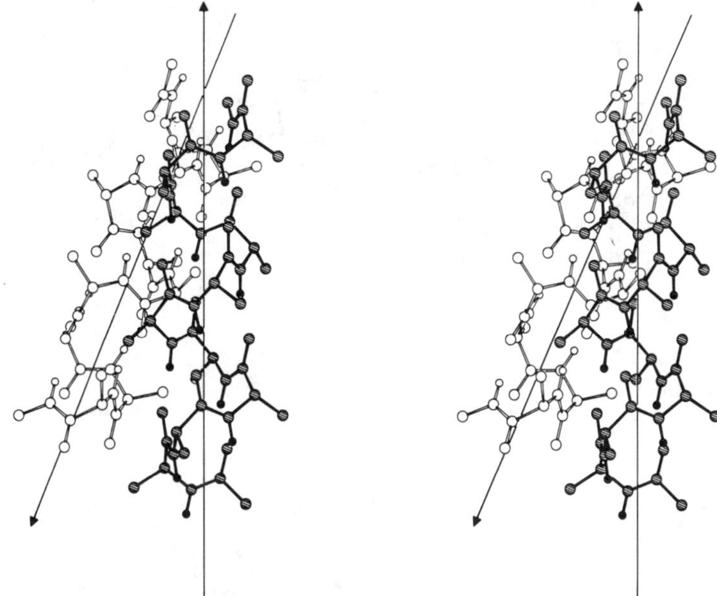

FIGURE 10. Stereoscopic picture of two CH_3CO-(L-Ala)$_{10}$-$NHCH_3$ α-helices in the lowest-energy packing state[41] (with a torsion angle of $-154°$). The helix axes are indicated by arrows, with the head of the arrow pointing in the direction of the C-terminals of each helix. [Reprinted with permission from ref. 41. Copyright 1983 American Chemical Society.]

decapeptide, and FIGURE 11B provides a diagram of the subsequently determined X-ray structure[46] which agrees[45,47] with the predicted one. When the structure was computed[44] in 1975, it was stated that "the ornithine side chain may be free to occupy more than one rotational state . . . (and there is) a hydrogen bond between a δ-NH_2 proton of Orn and the backbone CO of Phe. There is no experimental evidence indicating the existence of this interaction." When the X-ray structure was determined[46] in 1978, it was stated that "there is an intramolecular hydrogen bond . . . between . . . the ornithine side-chain nitrogen atom and the D-phenylalanine carbonyl oxygen atom which had not been predicted." FIGURE 11A, which has a computed rotational variant[45] of the previously computed[44] side-chain conformation of ornithine, clearly shows the predicted hydrogen bond (the original computed rotational variant[44] had the Orn and Phe partners interchanged).

The computed result of FIGURE 11A is not a fortuitously obtained one, but was arrived at by energy minimization from over 10,000 starting conformations. To provide an idea of the magnitude of the number of possible conformations, we refer to FIGURE 12, which shows some of the computed low-energy conformations of cyclo-hexaglycine.[48] These are conformations that have some kind of symmetry, and there are still others (obtained by a Monte Carlo procedure[49]) with nonsymmetric conformations. A decapep-

A

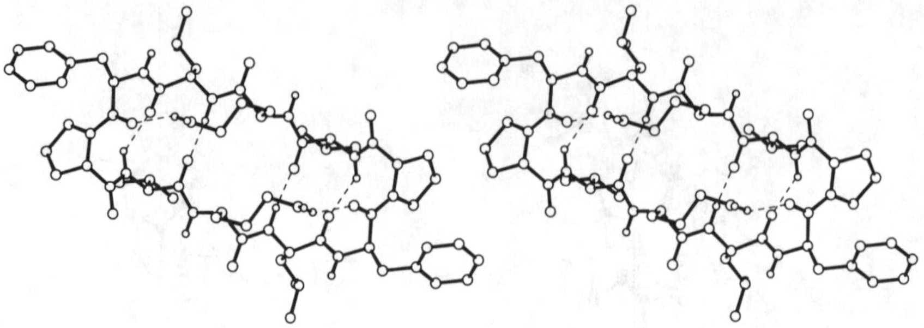

B

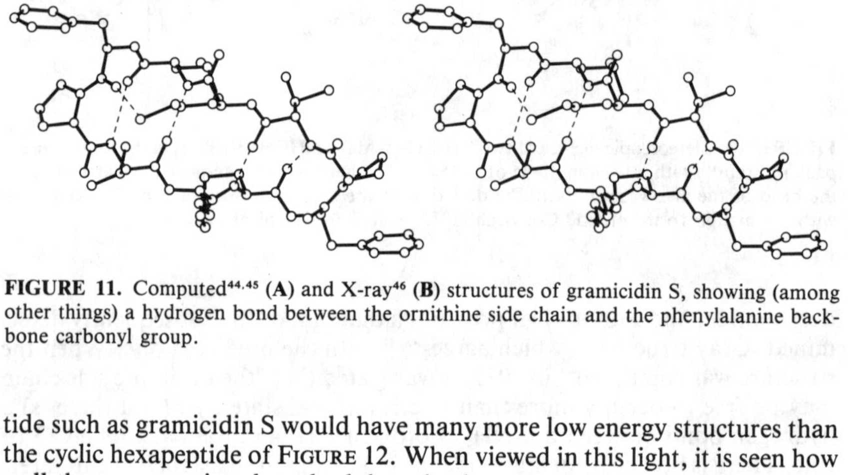

FIGURE 11. Computed[44,45] (**A**) and X-ray[46] (**B**) structures of gramicidin S, showing (among other things) a hydrogen bond between the ornithine side chain and the phenylalanine backbone carbonyl group.

tide such as gramicidin S would have many more low energy structures than the cyclic hexapeptide of FIGURE 12. When viewed in this light, it is seen how well the computational methodology leads to the fairly unique low-energy conformation of FIGURE 11A.

A larger oligopeptide, the 20-residue membrane-bound portion of melittin has also yielded to this methodology. Only two (very similar) low-energy, largely α-helical structures were found[50]; these are shown in FIGURE 13. X-ray[51] and NMR[52] structural information is available for melittin either as a tetramer in a crystal or as a monomer bound to micelles, respectively; considering possible environmental effects in either of these forms, the general agreement with experiment is satisfactory.

Turning to larger globular proteins, we defer until the next section the prediction of their three-dimensional structures, and consider in this section some related computations involving globular proteins. Two types of computations that overcome the multiple-minima problem to some extent are the energetic refinement of X-ray data and the computation of the structure

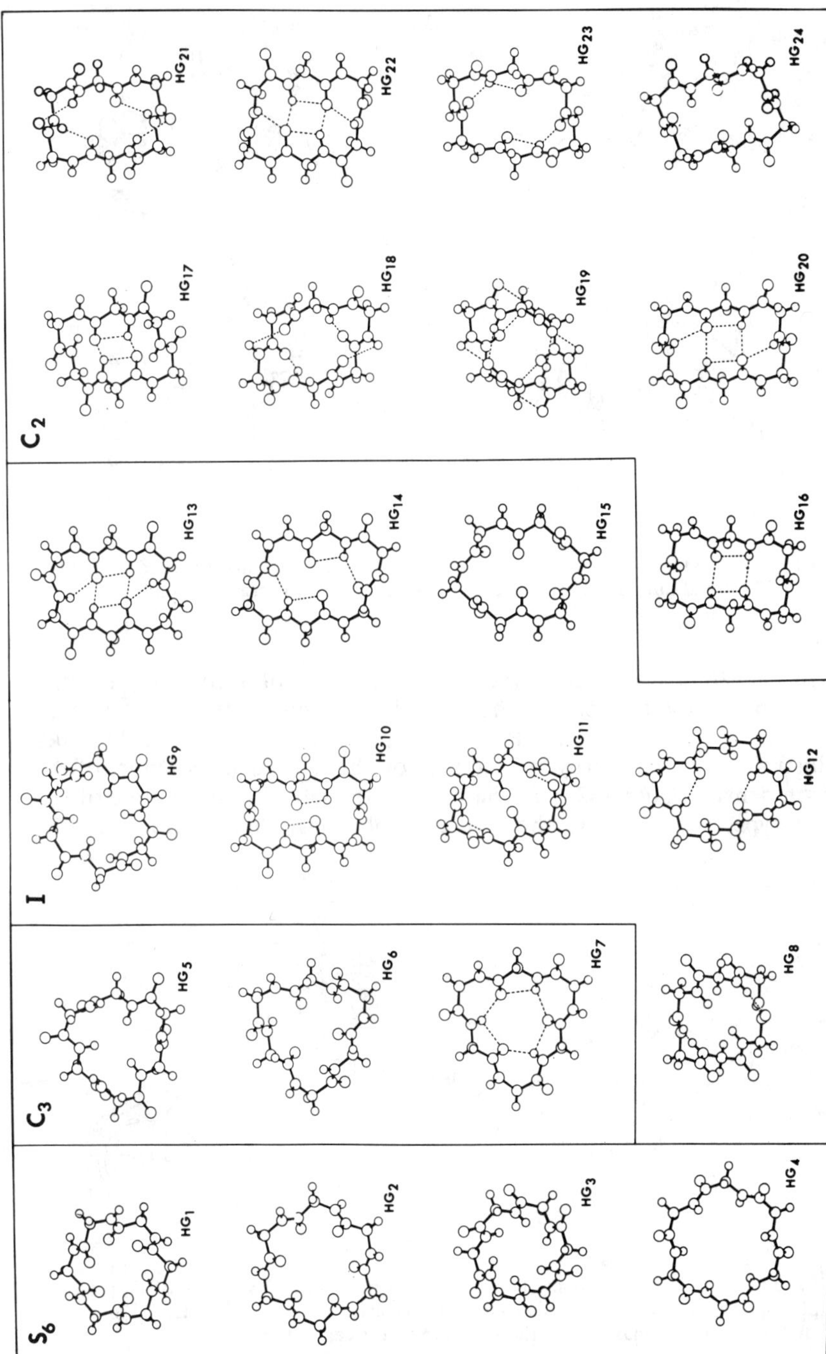

FIGURE 12. Computed low-energy structures of cyclic hexaglycine, with various kinds of symmetry as indicated.[48] [Reprinted with permission from ref. 48. Copyright 1973 American Chemical Society.]

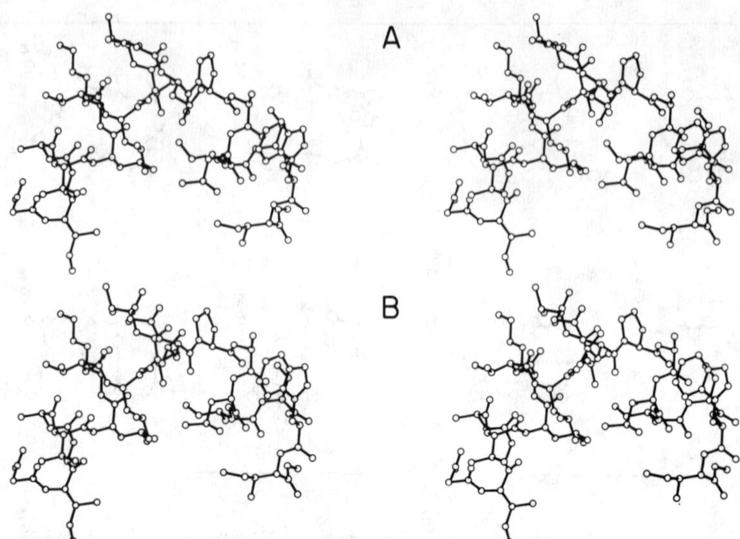

FIGURE 13. Stereo view of the two lowest energy structures calculated for residues 1–20 of melittin.[50] **(B)** global minimum; **(A)** structure of slightly higher energy (1.5 kcal/mol).

of a protein by making use of the known structure of a homologous one. An example of the first type is the potential energy-constrained real-space refinement of the structure of BPTI, using limited diffraction data.[53] FIGURE 14 illustrates how the energy-refined map on the right identifies more of the electron density of Arg-42 than the map on the left. As an example of the second type, we illustrate the structure of α-lactalbumin (FIGURE 15),

FIGURE 14. Comparison of sections of electron density maps around Arg-42 of BPTI, projected onto the xy-plane.[53] (*Left*) Map calculated by using the experimental 2.5 Å phases. (*Right*) Map calculated by using the phases from the structure factor calculation.

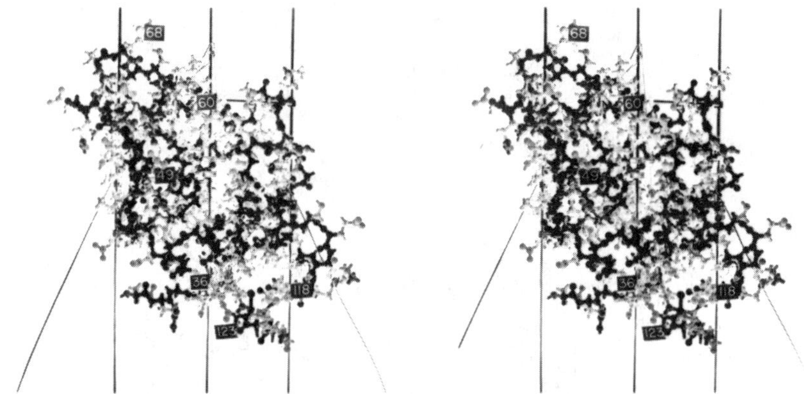

FIGURE 15. Stereo view of computed structure[54] of α-lactalbumin. [Reprinted with permission from ref. 54. Copyright 1974 American Chemical Society.]

computed[54] by using the known structure of hen egg white lysozyme. Although the X-ray structure of α-lactalbumin has not yet been determined, the predicted structure is at least in agreement with various experimental results cited in ref. 54 and some[55,56] that were obtained subsequently. Similar computations have been carried out[57] for several neurotoxins that are homologous to BPTI.

The same methodology is also applicable to intermolecular interactions, *e.g.*, enzyme-substrate complexes, thereby providing information about the structures and energetics involved in molecular recognition and specificity. Molecular details of such enzyme-substrate complexes are required in order to understand the mechanism of enzyme action; *i.e.*, simplified models (such as a virtual-bond chain) cannot provide such details. The computation pertains essentially to a docking process in which a flexible substrate approaches a flexible enzyme as the energy is minimized. Such calculations have been carried out for complexes of α-chymotrypsin with oligopeptides[58,59] and of hen egg white lysozyme with oligosaccharides.[59,60] FIGURE 16 illustrates the computed low-energy binding mode of a tripeptide to α-chymotrypsin, and FIGURE 17 shows the computed structure of the corresponding acyl-enzyme intermediate.

FIGURE 18 illustrates two computed binding modes[60] for hexasaccharide substrates of hen egg white lysozyme, a "left-sided" binding mode and a "right-sided" one. The former was predicted to predominate for (GlcNAc)$_6$. This prediction has recently been verified[61] by experiments involving competition between oligosaccharides and monoclonal antibodies for binding to hen egg white lysozyme and by measurements of the Michaelis-Menten constant K_M for hen egg white lysozyme and for a homologous lysozyme from ringed neck pheasant, the latter having several different amino acids in the "right-sided" binding site (see FIGURE 19).

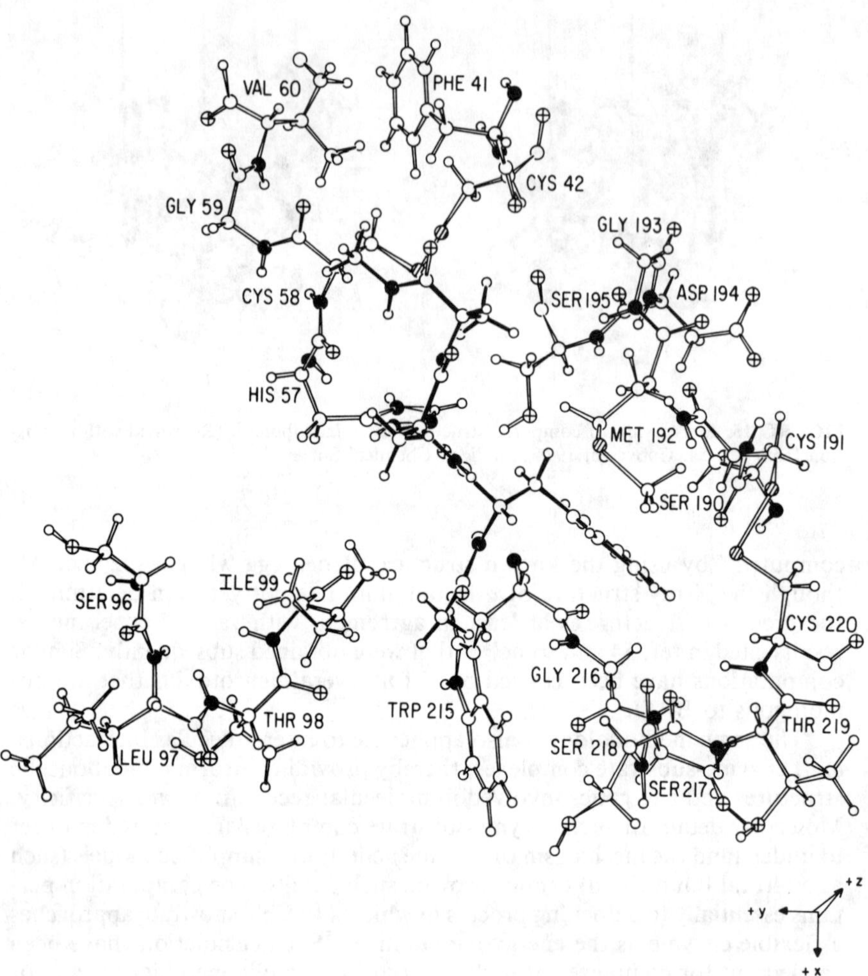

FIGURE 16. Lowest-energy conformer of N-acetyl-Phe-Ala-Ala-NH$_2$ in the cleft binding region of α-chymotrypsin.[59]

PREDICTION OF THREE-DIMENSIONAL STRUCTURES OF GLOBULAR PROTEINS

Our present procedures for computing the structures of globular proteins involve the build-up of energy-minimized fragments,[50,62-64] incorporation of short-, medium-, and long-range interactions,[65,66] and use of distance constraints[67] and empirical procedures for including hydration.[8,9] In the build-up procedure, collections of low-energy structures are retained, and larger structures are built up from these and their energies are then minimized.

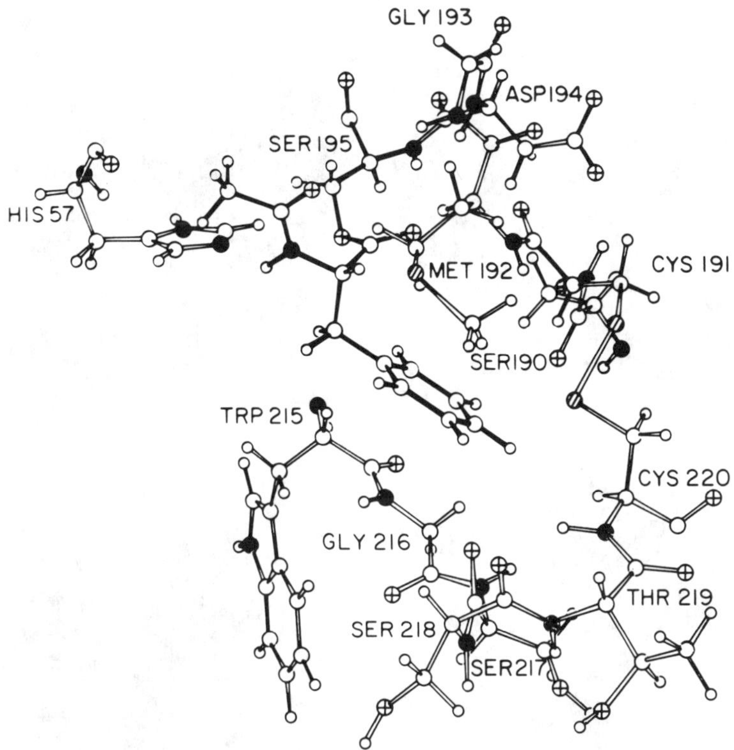

FIGURE 17. Lowest-energy conformation for the acyl-enzyme intermediate of chymotrypsin with *N*-acetyl-L-phenylalanine.[59]

FIGURE 20 illustrates the ordering of the energies of local conformations, and the discarding of higher-energy ones as the chain is built up from fragments,[62] as illustrated in FIGURE 21. It is important to note that, in the build-up procedure, there are many low-energy fragments that must be retained in the allowed set from which to build up longer structures; *i.e.,* premature discarding of some of these structures could lead to erroneous results at later stages of the computations. Further, at say the tetra- or pentapeptide stage, several different conformations can have very similar energies, and the "correct" conformation of such a small oligopeptide *in a globular protein* is selected as a low-energy one only at a later stage of the computations, when the fragments are much longer and their conformations are influenced by long-range (in addition to short- and medium-range) interactions.

Various short- and medium-range algorithms (partially illustrated in FIGURE 3) are used to predict *initial* α-helical, extended-structure, and β-bend conformations. Because this information is not sufficient to obtain the proper shape of the molecule, as illustrated in FIGURE 2, these initial assignments

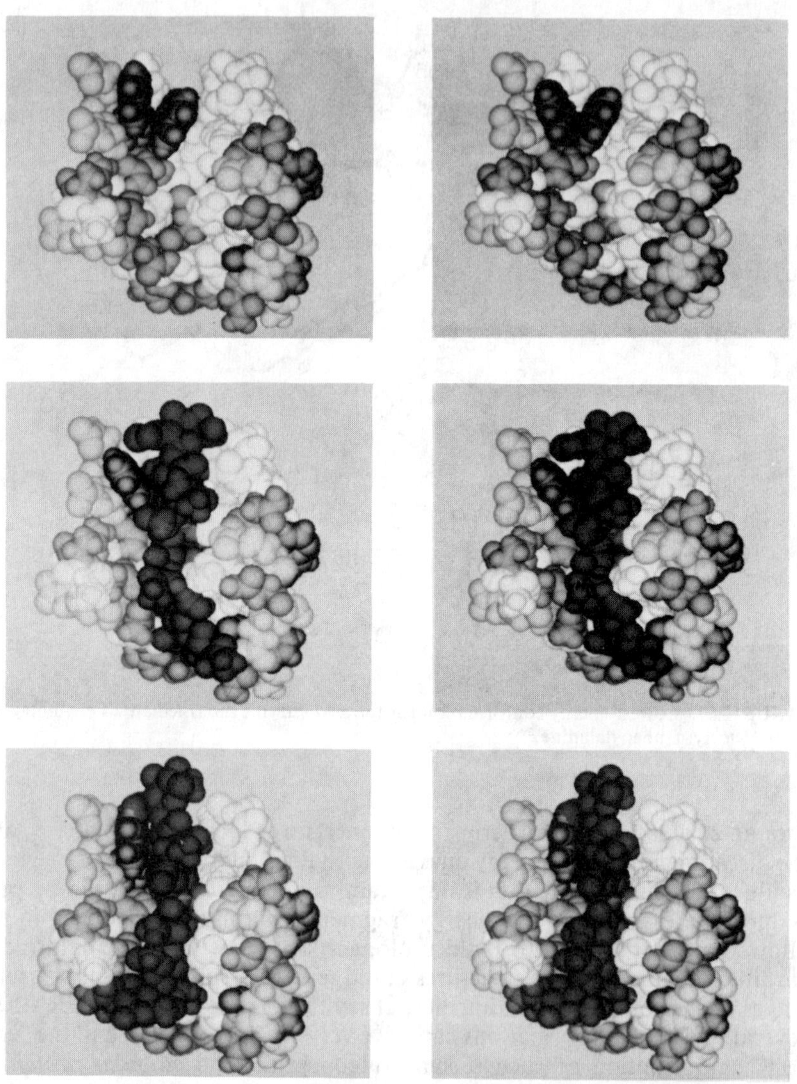

FIGURE 18. Stereo views of space-filling models[60] of (*top*) the active site of native hen egg white lysozyme; (*middle*) energy-minimized model-built hexamer [(GlcNAc)$_6$] bound to the active site (right-sided mode); and (*bottom*) lowest-energy hexamer bound to the active site (left-sided mode). [Reprinted with permission from ref. 60. Copyright 1979 American Chemical Society.]

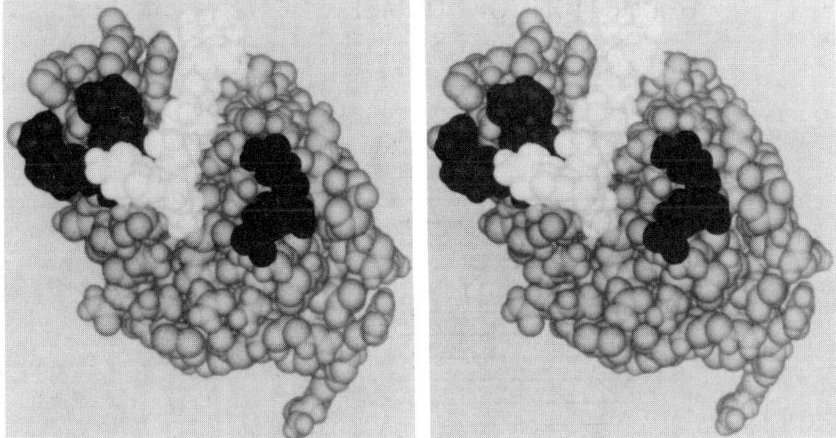

FIGURE 19. Stereo view[61] of the computed structure[60] of the low-energy complex between (GlcNAc)$_6$ and hen egg white lysozyme (same complex as in the bottom of Fig. 18). The dark shaded regions contain the residues involved in left- and right-sided binding, respectively. Even though hen and ringed neck pheasant lysozymes differ in the dark-shaded region on the right side, both exhibit similar values of K_M toward (GlcNAc)$_6$.

(and corrections for incorrect assignments) are improved during the subsequent constrained energy minimization.[66] Long-range interactions are incorporated initially by dividing the conformational space of the protein into groups, according to the spatial geometric arrangements of their loops (SGALs).[65] Such SGALs incorporate long-range pairwise interactions. Conformations are then selected randomly from each group of SGALs for energy minimization of the *full-atom* chain. The assumption that a structure selected from the group of SGALs in which the native structure lies would have a lower energy than one selected from another group has been verified in a preliminary way[65,66]; *i.e.*, the energy of structure R of FIGURE 22 is lower than that of structure W after minimization. This work is continuing by making use of interactive computer graphics to select conformations from the various groups of SGALs.

The conformations obtained by minimizing the energies of structures from the various groups of SGALs, along with conformations obtained by other criteria, are optimized[67] by incorporation of distance constraints and effects of hydration.

In the optimization computations, both real chains[65,66] and virtual-bond chains[67] are used. Recently, we developed a procedure[68] to convert a virtual-bond chain to a real one; this enables us to cycle back-and-forth between procedures that make use of the two types of chain representation. Thus far, the virtual-bond procedure has led to a structure of BPTI[67] with an rms deviation of 2.2 Å from the experimental one. By cycling between the various procedures, and ultimately carrying out a full-scale energy minimization *without approximations,* using an array processor,[1] we hope to be able to

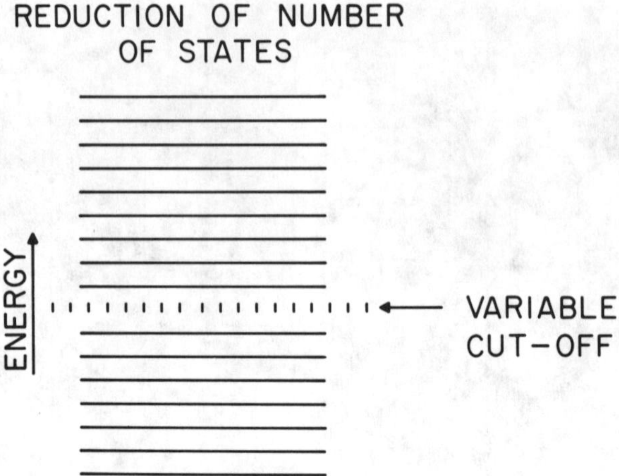

FIGURE 20. Schematic representation of the energies of the various minima in the conformational space of an oligopeptide. In using such a data base to build up larger structures, only those below a (variable) cut-off are considered. [Reprinted with permission from Biopolymers **22**: 1. Copyright 1983 John Wiley & Sons.]

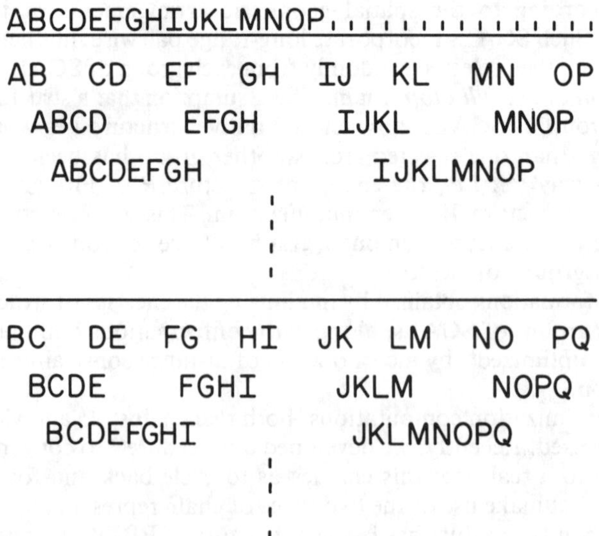

FIGURE 21. Schematic representation of the build-up of larger structures from smaller ones (AB, CD, . . ., ABCD, EFGH, . . . or, alternatively, BC, DE, . . ., BCDE, FGHI, . . .). [Reprinted with permission from Biopolymers **22**: 1. Copyright 1983 John Wiley & Sons.]

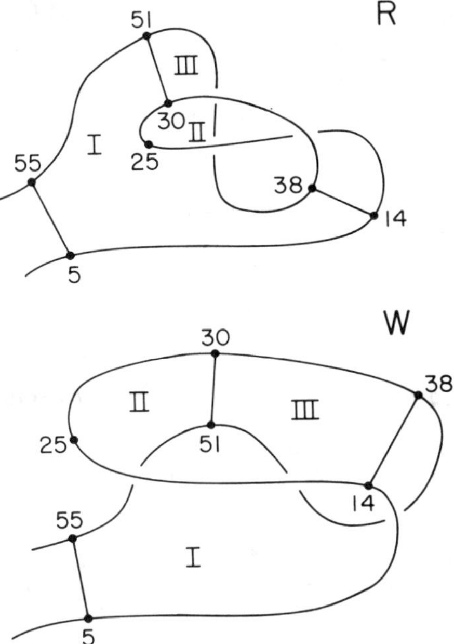

FIGURE 22. The three loops of BPTI. *R* and *W* are, respectively, the right and wrong spatial geometric arrangements of the loops.[65] [Reprinted with permission from ref. 65. Copyright 1981 American Chemical Society.]

improve the predictability for a globular protein having of the order of 100–200 residues.

The techniques developed for carrying out the computations on the systems described above are also applicable to design modified proteins that can be synthesized by currently available genetic engineering methodology. They are also applicable to designing drugs for binding to specific cell-surface receptors, and to probe the structures of such receptors.

FIGURE 23 is a schematic representation of the kinds of problems that can be treated by the methodology used here. Direct application to X-ray data leads to low-energy structures.[53,57,59] Such refined structures provide starting points to compute the conformations of homologous proteins,[54,57] which in turn can be used (in a docking process) to treat the binding of enzymes to substrates (the process characterized by K_M),[58-60,70] the binding of drugs and hormones to receptors, and the binding of proteins to nucleic acids. With sufficient experimental information about structure-activity relationships, this technique can be applied to map the active site of an unknown receptor. The same energy minimization algorithms can be used (together with constraints) in protein folding studies. Two potential extensions of this methodology are to intermolecular interactions in multi-subunit proteins such as hemoglobin, and to the structures and energetics of activation complexes (the process characterized by k_{cat}).[71] In the latter case, the computations

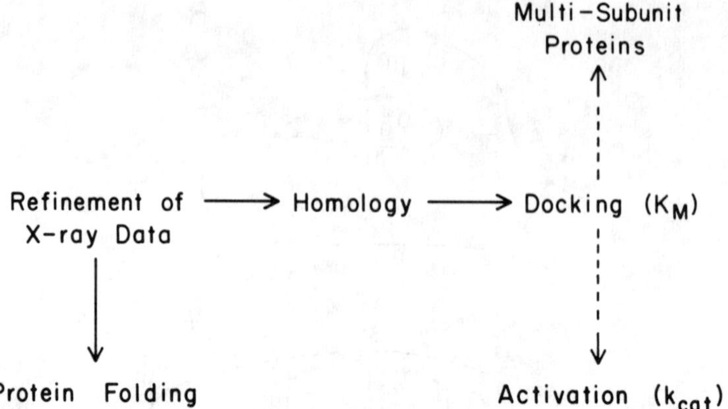

FIGURE 23. Various problems treated by energy minimization. The *solid arrows* pertain to computations already carried out; the *dashed arrows* pertain to potential new applications.

must allow for bond stretching and bond angle bending, in addition to the variation of dihedral angles that is allowed for in all of the other computations.

REFERENCES

1. POTTLE, C., M. S. POTTLE, R. W. TUTTLE, R. J. KINCH & H. A. SCHERAGA. 1980. J. Computational Chem. **1**: 46.
2. GIBSON, K. D. & H. A. SCHERAGA. 1969. Physiol. Chem. Phys. **1**: 109.
3. GIBSON, K. D. & H. A. SCHERAGA. 1969. Proc. Natl. Acad. Sci. U.S.A. **63**: 242.
4. GŌ, N. & H. A. SCHERAGA. 1969. J. Chem. Phys. **51**: 4751.
5. GŌ, N. & H. A. SCHERAGA. 1976. Macromolecules **9**: 535.
6. GŌ, N., P. N. LEWIS & H. A. SCHERAGA. 1970. Macromolecules **3**: 628.
7. HODES, Z. I., G. NÉMETHY & H. A. SCHERAGA. 1979. Biopolymers **18**: 1565.
8. MEIROVITCH, H. & H. A. SCHERAGA. 1980. Macromolecules **13**: 1406.
9. MEIROVITCH, H. & H. A. SCHERAGA. 1981. Macromolecules **14**: 340.
10. SCHERAGA, H. A. 1981. Biopolymers **20**: 1877.
11. SCHERAGA, H. A. 1983. Biopolymers **22**: 1.
12. SCHERAGA, H. A. 1967. Fed. Proc. **26**: 1380.
13. WLODAWER, A., R. BOTT & L. SJÖLIN. 1982. J. Biol. Chem. **257**: 1325.
14. LEACH. S. J., G. NÉMETHY & H. A. SCHERAGA. 1977. Biochem. Biophys. Res. Commun. **75**: 207.
15. BRAUN, W., C. BOSCH, L. R. BROWN, N. GŌ & K. WÜTHRICH. 1981. Biochim. Biophys. Acta **667**: 377.
16. C. A. MCWHERTER, E. HAAS & H. A. SCHERAGA. Work in progress.
17. NÉMETHY, G. & H. A. SCHERAGA. 1965. Biopolymers **3**: 155.
18. HAVEL, T. F., G. M. CRIPPEN & I. D. KUNTZ. 1979. Biopolymers **18**: 73.
19. WAKO, H. & H. A. SCHERAGA. 1981. Macromolecules **14**: 961.
20. BURGESS, A. W. & H. A. SCHERAGA. 1975. Proc. Natl. Acad. Sci. U.S.A. **72**: 1221.
21. WAKO, H., N. SAITÔ & H. A. SCHERAGA. 1983. J. Protein Chem. **2**: 221.

22. LEACH, S. J., G. NÉMETHY & H. A. SCHERAGA. 1966. Biopolymers **4**: 887.
23. NÉMETHY, G., D. C. PHILLIPS. S. J. LEACH & H. A. SCHERAGA. 1967. Nature **214**: 363.
24. YAN, J. F., F. A. MOMANY & H. A. SCHERAGA. 1970. J. Am. Chem. Soc. **92**: 1109.
25. ERENRICH, E. H., R. H. ANDREATTA & H. A. SCHERAGA. 1970. J. Am. Chem. Soc. **92**: 1116.
26. CHOU, K. C., M. POTTLE, G. NÉMETHY, Y. UEDA & H. A. SCHERAGA. 1982. J. Mol. Biol. **162**: 89.
27. CHOU, K. C. & H. A. SCHERAGA. 1982. Proc. Natl. Acad. Sci. U.S.A. **79**: 7047.
28. SCHERAGA, H. A., K. C. CHOU & G. NÉMETHY. 1982. *In* Conformation in Biology. R. Srinivasan & R. H. Sarma, Eds: 1. Adenine Press. Guilderland, New York.
29. CHOU, K. C., G. NÉMETHY & H. A. SCHERAGA. 1983. J. Mol. Biol. **168**: 389.
30. CHOU, K. C., G. NÉMETHY & H. A. SCHERAGA. 1983. Biochemistry. **22**: 6213.
31. CHOTHIA, C. 1973. J. Mol. Biol. **75**: 295.
32. GŌ, M., F. T. HESSELINK, N. GŌ & H. A. SCHERAGA. 1974. Macromolecules **7**: 459.
33. ALTER, J. E., R. H. ANDREATTA, G. T. TAYLOR & H. A. SCHERAGA. 1973. Macromolecules **6**: 564.
34. ZIMM, B. H. & J. K. BRAGG. 1959. J. Chem. Phys. **31**: 526.
35. MATTICE, W. L. & H. A. SCHERAGA. 1984. Biopolymers **23**: 1701.
36. MILLER, M. H. & H. A. SCHERAGA. 1976. J. Polymer Sci., Polymer Symp. **54**: 171.
37. OKUYAMA, K., N. TANAKA, T. ASHIDA & M. KAKUDO. 1976. Bull. Chem. Soc. Japan **49**: 1805.
38. MILLER, M. H., G. NÉMETHY & H. A. SCHERAGA. 1980. Macromolecules **13**: 470.
39. MILLER, M. H., G. NÉMETHY & H. A. SCHERAGA. 1980. Macromolecules **13**: 910.
40. NÉMETHY, G., M. H. MILLER & H. A. SCHERAGA. 1980. Macromolecules **13**: 914.
41. CHOU, K. C., G. NÉMETHY & H. A. SCHERAGA. 1983. J. Phys. Chem. **87**: 2869.
42. CRICK, F. H. C. 1953. Acta Crystallogr. **6**: 689.
43. NÉMETHY, G. 1983. Biopolymers **22**: 33.
44. DYGERT, M., N. GŌ & H. A. SCHERAGA. 1975. Macromolecules **8**: 750.
45. NÉMETHY, G. & H. A. SCHERAGA. 1984. Biochem. Biophys. Res. Commun. **118**: 643.
46. HULL, S. E., R. KARLSSON, P. MAIN, M. M. WOOLFSON & E. J. DODSON. 1978. Nature **275**: 206.
47. RACKOVSKY, S. & H. A. SCHERAGA. 1980. Proc. Natl. Acad. Sci. U.S.A. **77**: 6965.
48. GŌ, N. & H. A. SCHERAGA. 1973. Macromolecules **6**: 525.
49. GŌ, N. & H. A. SCHERAGA. 1978. Macromolecules **11**: 552.
50. PINCUS, M. R., R. D. KLAUSNER & H. A. SCHERAGA. 1982. Proc. Natl. Acad. Sci. U.S.A. **79**: 5107.
51. TERWILLIGER, T. C., L. WEISSMAN & D. EISENBERG. 1982. Biophys. J. **37**: 353.
52. BROWN. L. R., W. BRAUN, A. KUMAR & K. WÜTHRICH. 1982. Biophys. J. **37**: 319.
53. FITZWATER, S. & H. A. SCHERAGA. 1982. Proc. Natl. Acad. Sci. U.S.A. **79**: 2133.
54. WARME, P. K., F. A. MOMANY, S. V. RUMBALL, R. W. TUTTLE & H. A. SCHERAGA. 1974. Biochemistry **13**: 768.
55. BERLINER, L. J. & R. KAPTEIN. 1981. Biochemistry **20**: 799.
56. GERKEN, T. A. 1983. Fed. Proc. **42**: 2001.
57. SWENSON, M. K., A. W. BURGESS & H. A. SCHERAGA. 1978. *In* Frontiers in Physicochemical Biology. B. Pullman, Ed.: 115. Academic Press. New York.

58. PLATZER, K. E. B., F. A. MOMANY & H. A. SCHERAGA. 1972. Int. J. Peptide Protein Res. **4**: 201.
59. SCHERAGA, H. A., M. R. PINCUS & K. E. BURKE. 1982. *In* Structure of Complexes between Biopolymers and Low Molecular Weight Molecules. W. Bartmann & G. Snatzke, Eds.: 53. John Wiley. Chichester.
60. PINCUS, M. R. & H. A. SCHERAGA. 1979. Macromolecules **12**: 633.
61. SMITH-GILL, S. J., J. A. RUPLEY, M. R. PINCUS, R. P. CARTY & H. A. SCHERAGA. 1984. Biochemistry **23**: 993.
62. SCHERAGA, H. A. 1974. *In* Current Topics in Biochemistry, 1973. C. B. Anfinsen & A. N. Schechter, Eds: 1. Academic Press. New York.
63. SIMON, I., G. NÉMETHY & H. A. SCHERAGA. 1978. Macromolecules **11**: 797.
64. PINCUS, M. R. & R. D. KLAUSNER. 1982. Proc. Natl. Acad. Sci. U.S.A. **79**: 3413.
65. MEIROVITCH, H. & H. A. SCHERAGA. 1981. Macromolecules **14**: 1250.
66. MEIROVITCH, H. & H. A. SCHERAGA. 1981. Proc. Natl. Acad. Sci. U.S.A. **78**: 6584.
67. WAKO, H. & H. A. SCHERAGA. 1982. J. Protein Chem. **1**: 5, 85.
68. PURISIMA, E. O. & H. A. SCHERAGA. 1984. Biopolymers **23**: 1207.
69. WARME, P. K. & H. A. SCHERAGA. 1974. Biochemistry **13**: 757.
70. SCHERAGA, H. A. 1984. Pont. Acad. Sci. Ser. Var. **55**: 21.
71. PINCUS, M. R. & H. A. SCHERAGA. 1981. Acc. Chem. Res. **14**: 299.

A Machine Architecture for Molecular Dynamics: The Systolic Loop[a]

NEIL S. OSTLUND

Department of Computer Science
University of Waterloo
Waterloo, Ontario N2L 3G1

ROBERT A. WHITESIDE[b]

Departments of Computer Science and Chemistry
Carnegie-Mellon University
Pittsburgh, Pennsylvania 15213

INTRODUCTION

The method of molecular dynamics (MD) is an increasingly important computation tool in theoretical studies of many molecular systems, ranging from simple liquids and solutions to biopolymers.[1] This technique involves the direct simulation of molecular motion by a time-step by time-step integration of the equations of motion of the system. In many cases, however, the applicability of this technique is severely limited by available computing resources. For instance, MD treatments of rather modest-sized proteins currently run for less than 1 ns of simulated time.[2,3] However, much interesting biochemistry occurs on microsecond or even millisecond time scales. Large scale parallel computer architectures, in which hundreds or even thousands of processing elements participate in a computation, offer one hope of providing the processing power necessary to help "solve" some of these problems. However, this will only be possible if an efficient parallel algorithm and a matching parallel architecture are known for the problem. Present trends suggest that special purpose architectures may be the best hope for such calculations.

Many designs for highly parallel systems have the processing elements arranged in some kind of regular structure (a line, a square, or hexagonal array, for instance) in which direct interprocessor communication is possible only between nearest neighbors. The modularity and regularity in such a system makes it very attractive from a hardware point of view. This is true

[a] Funds for this research were provided by the National Science Foundation under NSF grant no. MSC–79–20698, and from the Natural Sciences and Engineering Research Council of Canada under NSERC grant no. A2490 (1983–84). The Intel Corporation, via Emil Sarpa and Tony Anderson, has generously donated equipment for this research.
[b] Present address: Sandia National Laboratory, Livermore, CA 94550

not only for VLSI systems, in which many specialized processors may be present on a single chip, but also for more general-purpose multiprocessors in which the processing elements are conventional computers. However, the utility of such a machine depends upon the range of problems that have efficient implementations onto the hardware. One obvious requirement is that the communication requirements of the parallel algorithm map efficiently onto the nearest-neighbor interprocessor connections of the machine.

An important step forward in this area was made with the development of a class of algorithms termed *systolic* algorithms.[4,5] In a step (or pulse) of a systolic algorithm, each processor exchanges some data values and/or partial results with its neighbors, then performs some simple computations, perhaps modifying the results. Thus, data and results pulse regularly through the machine, and the name "systolic" is an anaology to the flow of blood through the circulatory system. Depending upon the processor interconnection structure and the operations performed by each processor, a number of useful computations can be performed. Although the original motivation for these algorithms was efficient use of parallelism in VLSI systems, many of the same issues apply to multicomputer architectures also.

In previous papers,[6,7] we have described some systolic implementations of the Metropolis Monte Carlo algorithm for statistical mechanical simulations. This systolic parallel algorithm uses a very simple multiple processor architecture that we have called a "systolic loop." A systolic loop has processors arranged in a circular configuration so that each need communicate directly only with its two nearest neighbors in the loop. In the present paper, we present a systolic loop algorithm for molecular dynamics simulations and describe an implementation of the systolic loop architecture (WATERLOOP/2) that is being constructed at the University of Waterloo.

The remainder of the paper is structured as follows. First we briefly describe the systolic loop architecture. The next section presents the (serial) molecular dynamics algorithm of Verlet.[8] Following this, parallel systolic versions of Verlet's algorithm and a machine to execute these parallel algorithms are described. Finally, we describe the present status of this work and directions for the future.

THE SYSTOLIC LOOP ARCHITECTURE

As the density of circuitry that can be integrated onto the surface of a silicon chip increases and the cost of computer hardware goes down, new options are presented to the computer architect. The original systolic array, proposed by Kung and Leiserson,[4] is characterized by a one- or two-dimensional array of simple computational cells arranged in a regular (*e.g.*, rectangular or hexagonal) pattern. A calculation proceeds by having data flow into, through, and out of the array with the transfer of data between nearest neighbors controlled by a single master clock. The flow of data through the array of cells at regular ticks of the clock is analogous to the flow of blood through the arteries at regular pulses of the heart. These ideas attracted immediate attention because of the possibility of designing a single specialized chip and then obtaining massive parallelism by simple replication of the chip in one or two dimensions.

Systolic processing is thus characterized by a regular flow of data through a set of identical computational elements. It appears to avoid many of the communication problems characterized by the fetch-execute style of Von Neumann machines. The synchronization and contention problems of thousands of processors trying to access the same shared memory are likely to be overwhelming, whereas the nearest-neighbor-only communication pattern of a systolic system holds hope for using effectively a large number of processors in the solution of a single task.

The usual systolic array has cells that perform relatively simple operations, such as an inner product, with data moving between cells on the pulse of a global clock. The general algorithmic ideas, however, carry over to cells that are capable of general purpose computation and to an interprocessor transfer of data that, rather than being clocked (*i.e.,* synchronous), occurs via an agreement of readiness on the part of both a producer cell and consumer cell (*i.e.,* it is asynchronous). The usual systolic array has data flowing in one side of the array with results flowing out the other side of the array. For example, two matrices can flow into a hexagonal array and a matrix product flows out. Such computations appear to be particularly applicable to signal processing applications.

Large scientific calculations, on the other hand, usually involve repetition or iteration. This suggests a cyclic topology rather than the acyclic topology of the usual systolic array. The above two considerations have led to an architecture that we call a systolic loop. It is systolic in the sense that it has data flowing through a regular arrangement of identical cells. It differs from the systolic array of Kung and co-workers,[5] however, in the following two aspects: (1) the systolic loop has cells that are asynchronous and capable of general purpose stored program computation. Cells communicate via a producer-consumer relation rather than on the pulse of a clock, and (2) the systolic loop has data flowing around and around a loop (of systolic cells) in some form of repetitive or iterative computation, not in one side and out the other side of an array in a single pass.

FIGURE 1 shows in qualitative form a systolic loop. The flow of data is unidirectional. Each cell consists of a processor and its associated memory. The interprocessor communication mechanism is unspecified. It might consist of a first-in, first-out (FIFO) queue, a dual-ported shared memory, *etc.* The producer-consumer relation usually has one processor filling a buffer and another processor simultaneously emptying it. Any particular cell in the systolic loop is a producer for its consumer downstream and a consumer for its producer upstream. Every cell is identical in the sense that each executes the same code. Data-dependent branches are allowed, however, since the processors are asynchronous.

A simple example of a systolic loop algorithm is the matrix multiplication, $C = AB$. We divide all square matrices into columns and have each of the processors responsible for its particular subset columns of each of the

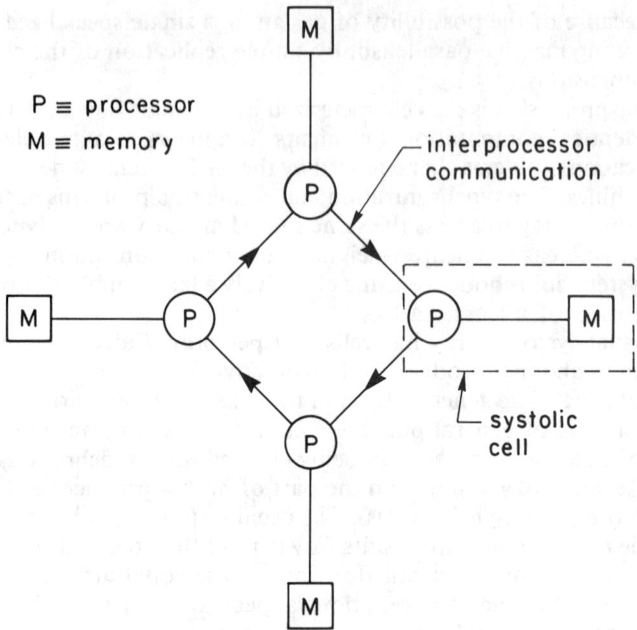

FIGURE 1. A systolic loop.

matrices. Processor i is responsible for columns $j+1, \ldots j+m$ of all matrices, for example. To calculate matrix $\mathbf{C}$, each processor sends off to its consumer its own columns of $\mathbf{A}$ and then receives the other columns of $\mathbf{A}$ in turn from its producer. It passes these columns on as well as using them in an outer product to build up locally the values of its own columns of $\mathbf{C}$. This continues until one's own columns of $\mathbf{A}$ come back (a boomerang effect) and the computation is complete.

THE (SERIAL) VERLET
MOLECULAR DYNAMICS ALGORITHM

The method of molecular dynamics consists of the step-by-step numerical integration of the equations of motion of a molecular system. The Verlet algorithm[8] is basically the finite difference form of Newton's equations. First, an initial set of positions and velocities for the particles is determined. For the simulation of a liquid, for instance, the particles might be placed at regularly spaced lattice sites and given random velocities based on a Maxwellian distribution for temperature T. A protein dynamics simulation might start with an experimental crystal structure and velocities of zero. However this initialization is done, given these coordinates for the particles to time $t_0 - \delta t$, the velocities can be used to predict a set of coordinates at a later time t_0. These two sets of coordinates (at times $t_0 - \delta t$ and t_0) represent the initialization of the system. The Verlet algorithm then gives the coordinates for all future times as:

$$\mathbf{r}_i(t + \delta t) = -\mathbf{r}_i(t - \delta t) + 2\mathbf{r}_i(t) + \mathbf{F}_i(t)/m_i(\delta t)^2 \ ,i = 1 \cdots n. \qquad (1)$$

In this equation, $\mathbf{r}_i(t)$ represents the coordinates of the ith particle at time t. $\mathbf{F}_i$ is the total force exerted on particle i of mass m_i. A usual approximation is that the total force exerted on a particle, $\mathbf{F}_i$, is given by a sum of pair-wise additive forces exerted by each of the other n–1 particles:

$$\mathbf{F}_i = \sum_{j \neq i}^{n} \mathbf{f}_{ij}(\mathbf{r}_i,\mathbf{r}_j) \ ,i = 1 \cdots n. \qquad (2)$$

Here, $\mathbf{f}_{ij}(\mathbf{r}_i,\mathbf{r}_j)$ is the force exerted on particle i by particle j.

The details of the serial Verlet molecular dynamics algorithm, in an algol-like language, are illustrated below:

```
comment: serial Verlet algorithm;
comment: the array rt holds the current coordinates of each particle, and the
array rtm holds the coordinates for the previous time-step;
while true do
   for i from 1 to n do
      force[i]: = 0;
      for j from 1 to n do
         if (i≠j) then
            force[i] +:= compute_force(rt[i],rt[j])
         fi
      od
   od;
   for i from 1 to n do
      rtp: = -rtm[i] + 2× rt(i) + force[i]/(mass[i]× (δt)²);
      rtm[i]: = rt[i];
      rt[i]: = rtp;
   od;
od;
```

The total force exerted on each particle is computed and stored into the array *force*, then the coordinates of each particle are advanced one time-step. Most of the computing time in this program is spent in the function *compute_force* which computes the pair-wise force exerted on one particle by another. Considerable savings could therefore be achieved by noting that the matrix of pair-wise forces is anti-symmetric. In **(2)** it is true that:

$$\mathbf{f}_{ij}(\mathbf{r}_i,\mathbf{r}_j) = -\mathbf{f}_{ji}(\mathbf{r}_j,\mathbf{r}_i). \qquad (3)$$

Thus, a version of the above algorithm that made use of this fact could run almost a factor of two faster. However, we present the algorithm first in this simplest form because it represents a good starting point for a parallel implementation.

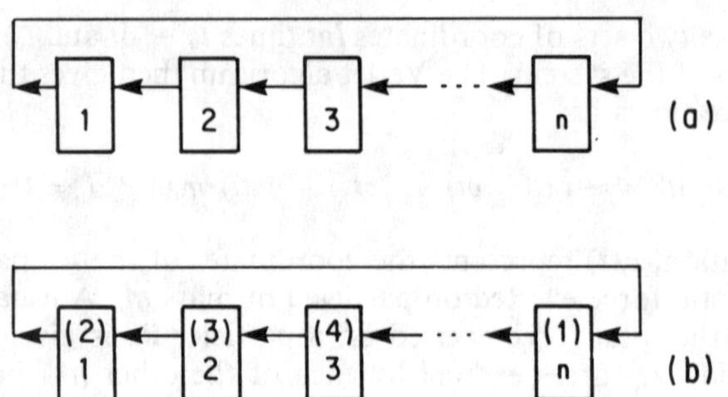

FIGURE 2. Flow of data in a parallel systolic loop molecular dynamics algorithm. The boxes represent processors that each have the coordinates of one particle, indicated by the numbers in the boxes **(a)**. During the force computation, the coordinates are passed around the loop in data packets illustrated by the parenthesized numbers **(b)**.

PARALLEL ALGORITHMS FOR MD

A parallel version of the above serial algorithm can be devised as follows: each of the n particles in the simulation is initially assigned one of n processors as illustrated in FIGURE 2a. These n processors are arranged as if in a loop in which each processor can communicate directly only with its two nearest neighbors. Each processor has the initialization coordinates, those at times t_0 and at $t_0 - \delta t$, for its assigned particle. As the simulation runs, each processor must at each step, (1) evaluate the total force exerted upon its assigned particle at time t, (2) use this to predict the coordinates of its particle at time $t + \delta t$, and (3) advance time to $t = t + \delta t$ and begin the next step. To evaluate the total force exerted on its assigned particle, each processor must obtain the current coordinates of all of the other n–1 particles in the system. To implement this step on the systolic loop, each processor first sends a copy of the current coordinates of its assigned particle to its left neighbor in the loop (FIGURE 2b). Each processor also, of course, receives coordinates from its right neighbor. Thus, each processor now holds the coordinates of two particles: the particle just received in a message (pkt_atom), and its local, assigned particle (local_atom). Each processor then: (1) sums into an accumulating total force for local_atom the contribution from pkt_atom, and (2) sends pkt_atom to its left neighbor, and receives a new pkt_atom from its right neighbor. This computation and data transfer constitute one *pulse* of the system. After n–1 such pulses, every processor has obtained and used the coordinates of every particle and, therefore, the correct total forces have been accumulated. Each processor can then predict new coordinates for its assigned particle, advance the time-step counter, and begin a new force computation. The details of the operations performed by a processor in the loop are represented below:

```
        comment: A parallel MD algorithm;
        while true do
            send(local_atom,local_x);
            force: = 0;
            for i from 1 to n − 1 do
                receive(pkt_atom,pkt_x);
                force + : =  compute_force(local_x,pkt_x);
                send(pkt_atom,pkt_x);
            od;
            receive(pkt_atom,pkt_x);
            new_x: =  predictor(force,local_x,former_x);
            time: = time  +  δt;
            former_x: = local_x;
            local_x: =  new_x;
        od;
```

Processors communicate via the message-passing primitives *send* and *receive*. Although they are not used above, the sequential indices of the atoms are passed around in the messages as *pkt_atom* for compatibility with the next algorithm. It is not necessary to have exactly n processors in the loop, although the algorithm presented requires this. If, as is likely the case, there are many more atoms in the simulated system than available processors, it is a simple matter to have a single processor perform the operations of several algorithmic processors (*processes*).

An Improvement

The MD simulation can be speeded up by taking advantage of the antisymmetry of the matrix of pair-wise forces. The algorithms presented so far, however, compute both $\mathbf{f}_{ij}$ and $\mathbf{f}_{ji}$. A serial algorithm avoiding this recomputation is presented below:

```
comment: The serial Verlet algorithm that avoids recomputation of pair-wise
force contributions;
while true do
    for i from 1 to n − 1 do
        force[i]: = 0;
        for j from i + 1 to n do
            temp := compute_force(rt[i],rt[j]);
            force[i] + : =  temp;
            force[j] − : =  temp;
        od;
    od;
    for i from 1 to n do
        rtp: = − rtm[i] + 2 × rt[i] + force [i]/(mass[i] × (δt)²);
        rtm[i]: = rt[i];
        rt[i]: = rtp;
    od;
od;
```

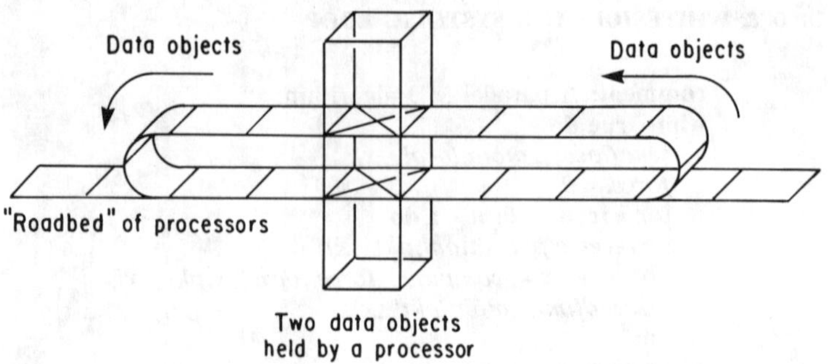

FIGURE 3. Illustration of the "tractor tread" flow of data along a "roadbed" of processors.

This scheme evaluates only the force contributions $\mathbf{f}_{ij}$ for $(j > i)$, producing a program that runs about twice as fast as the previous one. In practice, all serial MD programs take advantage of this anti-symmetry, and we now describe how this savings can be obtained in the parallel implementation.

Our MD problem is a member of a class of iterative n^2 problems in which, at each iteration, each of a set of n objects must interact computationally with each of the other $n-1$ objects. In the molecular dynamics case, of course, the n objects are the particles in the system being simulated. We have previously described[9] a general procedure to obtain efficient parallel implementations for problems of this class. Furthermore, this scheme has successfully been applied to the Metropolis Monte Carlo algorithm[6,7] for the simulation of molecular motion, as well as to Jacobi diagonalization[9] to obtain parallel algorithms. In this scheme, each data object and all of its associated state is logically organized into a data packet that can either be held stationary in a processor or be passed between processors in a message. Next, these data packets are arranged as if on a Caterpillar tractor tread that is lying on a "roadbed" of processors. This tractor analogy, illustrated in FIGURE 3, indicates the flow of data packets between the processors as it "rolls" along the processor roadbed. At any given time, each processor holds two data packets: the packet on the lower, stationary portion of the tractor tread on the roadbed at the processor's location, and the packet directly above the processor's location on the upper, moving part of the tractor tread. The general idea is that each processor performs (in parallel with the others) that part of the computation corresponding to the two data packets it currently holds. Next the tractor tread is advanced one step. Apart from complications at the ends of the tread, this "advancement" means computationally that each processor sends its data packet on the upper, moving part of the tractor tread to its left neighbor, and receives a new one from its right neighbor. Each processor then performs computations relating to this new pair of objects, and so on. After $n-1$ such pulses of the system, each object has met all of the other objects and one iteration of the computation is completed.

Although FIGURE 3 indicates a requirement for an infinite line of processors, it is easily seen that every two pulses a new processor on the left is activated by the receipt of its first data packet, while another processor at the

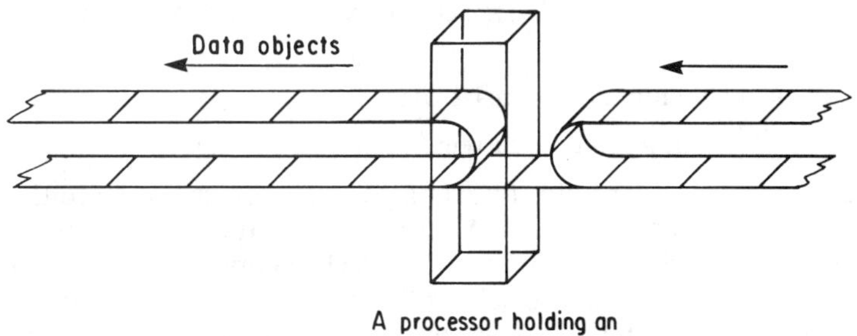

FIGURE 4. The flow of data around the systolic loop. The idea here is that data flowing off the edge reappears at the right.

right becomes idle because the tread has rolled completely past it. It is advantageous, therefore, to let the line of processors wrap around as illustrated in FIGURE 4. This forms a loop enabling the reuse of an idle processor rather than the activations of a new one. Furthermore, it is not necessary to have exactly $n/2$ processors to solve an n-object problem. It is a simple matter to have one physical processor simulate the activity of several adjacent algorithmic processors (processes).

The code below details the operations performed by a processor in the loop for the tractor tread MD algorithm:

```
comment: The parallel systolic loop MD algorithm avoiding the recomputa-
tion of interaction energies;
while true do
    local_force: = pkt_force: = 0;
    for pulse from 1 to n do
        send(pkt_atom,pkt_x,pkt_force);
        if (pkt_atom = (local_atom mod n) + 1) then
            pkt_atom: = local_atom;
            pkt_x: = local_x;
            pkt_force: = local_force;
            receive(local_atom,local_x,local_force)
        else
            receive(pkt_atom,pkt_x,pkt_force);
            t: = compute_force(local_x,pkt_x);
            local_force + : = t;
            pkt_force - : = t
        fi
    od;
    new_x: = predictor(local_force,local_x,local_former_x);
    local_former_x: = local_x;
    local_x: = new_x;
    new_x: = predictor(pkt_force,pkt_x,pkt_former_x);
    pkt_former_x: = pkt_x;
    pkt_x: = new_x;
    time: = time + δt;
od;
```

The data packets employed consist of:

$$\{\text{atom\_index, current\_}x, \text{former\_}x, \text{force}\}.$$

Initially, each processor holds two data packets corresponding to the lower and upper portions of the tread. After initializing the forces in its two data packets to zero, each processor repetitively: (1) computes the pair-wise force between its current pair of particles, (2) sums this contribution into the accumulating forces in both data packets, and (3) advances the tractor tread flow of data. Maintaining the correct pattern of data flow (especially at the two ends of the tractor tread) produces the principal new complication.

Normally, for a given processor, advancing the tread corresponds simply to sending its packet on the upper part and receiving a new one. However, when the processor holds the ends of the tread its operations must be different. The code above detects the end of the tread by assuming that the data packets are received sequentially in descending order. Thus when a processor's pkt_atom is one greater than its local_atom (modulo n), then it holds the trailing end of the tread, as is illustrated for the central processor in FIGURE 5a. At this point, the processor must send both of its packets to the left, and receive two packets, corresponding to the leading edge of the tread. Note that in FIGURE 5, and in the code above that implements this pattern, the processor dealing with the two ends of the tread spends a pulse idle—it performs no force computation, and produces a degradation of the performance of the parallel algorithm. However, this penalty is constant. One processor pulse is lost every two pulses independent of the problem size. Thus, the larger the number of particles in the simulation, the less significant this overhead will be.

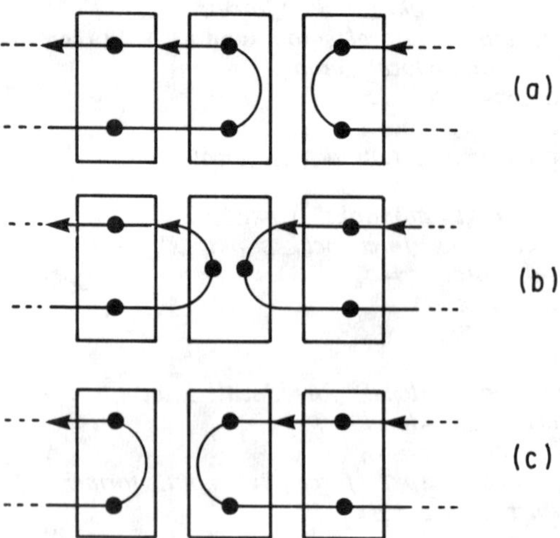

FIGURE 5. Sequence of events near the ends of the tractor tread flow of data. The boxes represent processors, and the circular blots represent data packets. The center processor is forced to be idle during pulse **b**.

The same techniques described here are applicable to other integration methods as well.[10]

A SYSTOLIC LOOP IMPLEMENTATION: WATERLOOP/2

A 64-processor implementation of the systolic loop architecture is underway. The basic architecture is sufficiently simple and modular that it is a relatively straightforward matter to put together prototypes of the architecture, particularly if they are bottom-up designs that use existing processors connected together in a simple fashion that does not require redesign of the processor.

In particular, many manufacturers produce single board computers. These single board computers normally have "parallel ports." These are input/output ports for communicating with peripherals, external devices being controlled by the board, *etc.* Communication is normally a byte at a time. A simple way of implementing a systolic loop thus consists of just connecting the parallel ports of a number of single board computers. Each board must have at least two ports, one input port and one output port. The input port of each board is connected to the output port of its producer and the output port of each board is connected to the input port of its consumer.

Our first implementation of the architecture (WATERLOOP/1) consists of three National Semiconductor Corporation DB16000 boards. These contain one of the new NS16032 32-bit microprocessors and this three-processor system is being used to explore a future design based on this microprocessor.

The second implementation (WATERLOOP/2) is a prototype that we will use to explore numerical applications, such as molecular dynamics. WATERLOOP/2 consists of 64 Intel iSBC 86/12A boards with an 8087 floating point coprocessor (iSBC 337) and an extra 32 Kbytes of memory (iSBC 330). Each processor thus has 64 Kbytes of local random access memory (RAM) for a total of 4 Mbytes of system RAM. In addition, each processor has 16 Kbytes of read-only memory (ROM). For 64-bit floating point operations, each processor has approximately one-sixth the speed of a VAX-11/780, so that the total system has the potential for being an order of magnitude faster than the VAX-11/780.

The processor cards are packaged as follows. The system consists of two 19-inch racks, each rack containing two 150 A (5 V) power supplies and two Multibus card cages. These two are thermally isolated so that the system consists, in essence, of four identical clusters. Each cluster (each Multibus card cage) holds 16 boards. Because of the dual-ported memory on each board, it is possible to share memory within a cluster. For performance reasons, this facility is not used for the systolic loop architecture and instead the parallel ports of the cards are connected together to form one large loop, as described above. However, we have allocated one board in each cluster to be outside the loop and to be a debugger module. The debugger has access to the memory of all cards in the same cluster via the shared memory. It can also start, stop, reset, and interrupt each of the boards in its cluster. The debugger thus controls each of the processors in its cluster.

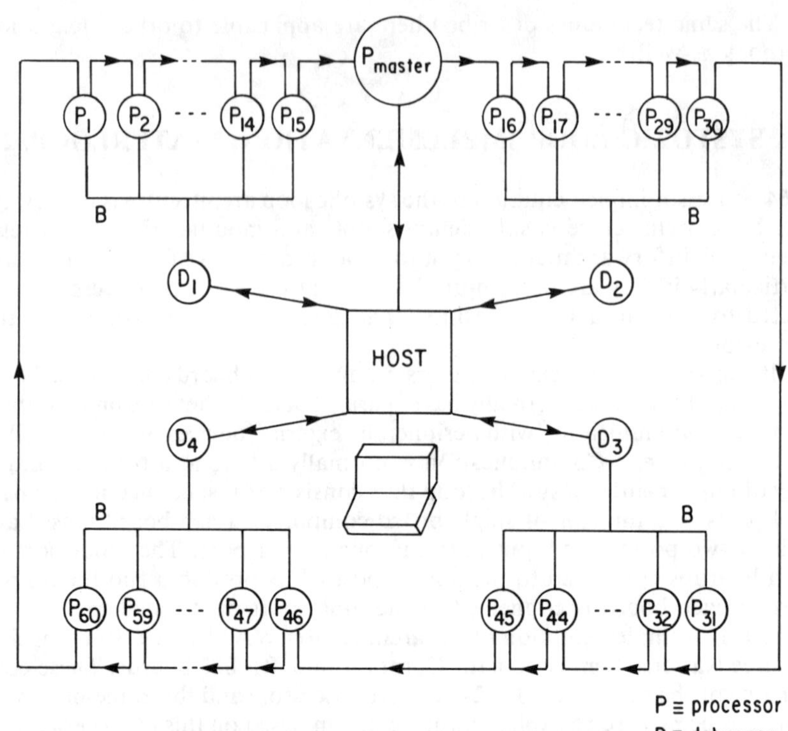

FIGURE 6. Pictorial view of WATERLOOP/2.

Outside the two racks sits a host processor. This host, an Intel 86/330 system with a 35 Mbyte hard disk, RMX-86 operating system, *etc.*, is used for compilation of programs, file storage, holding of results, *etc.* It interacts with the debugger modules for downloading of code, initializing the loop, collection of results, *etc.* The system is programmable in FORTRAN. Each processor holds an identical copy of ordinary serial code. The send and receive primitives that are required are implemented by simply writing to a special file (unit 0, for example) and reading from a special file (unit 1, for example).

FIGURE 6 shows a pictorial view of the system. The master processor is responsible for communicating with the host. Construction is still continuing, but as of 1 October 1983, the first of two racks had been powered-up and hardware construction is expected to be complete prior to the end of the year.

CONCLUSION

This systolic parallel algorithm for molecular dynamics simulations is both efficient and practical. The algorithm has been programmed and implemented[10] on a shared-memory multiprocessor (Cm*) and the systolic algorithm performs as well as shared-memory algorithms even though the Cm* architecture[11] is very different from the optimal systolic loop. WATERLOOP/2 will be functional shortly and will provide an excellent test bed for exploring systolic loop algorithms, including molecular dynamics calculations on biological systems. With experience gained from WATERLOOP/1 and WATERLOOP/2, a top-down design of a new system, WATERLOOP/3, will begin. This new systolic loop will be the first implementation really "designed" rather than "assembled" and can be expected to have much faster interprocessor communication hardware so as to be applicable to a wider range of problems.

SUMMARY

A multiprocessor algorithm for molecular dynamics calculations has been described. The new parallel algorithm, which corresponds to the serial Verlet algorithm, appears (in combination with a matching parallel architecture) to provide an effective way of applying a large number of processors to the very demanding computational requirements of molecular dynamics. The new architecture, called a systolic loop, has been presented and an implementation of the architecture, called WATERLOOP/2, has been described. This 64-processor system, which is nearing completion, is a simple design that uses the parallel ports of single board 8086/8087 microcomputers for interprocessor communication. For molecular dynamics calculations there is the expectation that the 64 processors will perform at approximately 64 times the speed of a single processor, about 10 times that of a VAX-11/780 in this particular instance.

REFERENCES

1. McCammon, J.A. & M. Karplus. 1980. Simulation of protein dynamics. Ann. Rev. Phys. Chem. **31**: 29–62.
2. McCammon, J. A., P. G. Wolynes & M. Karplus. 1979. Picosecond dynamics of tyrosine side chains in proteins. Biochemistry **18**: 927–942.
3. McCammon, J. A., B. R. Gelin & M. Karplus. 1977. Dynamics of folded proteins. Nature **267**(5612): 585–590.
4. Kung, H. T. & C. E. Leiserson. 1980. Systolic arrays (for VLSI). *In* Introduction to VLSI Systems. C. A. Mead & L. A. Conway, Eds. Addison-Wesley. Reading, Mass.
5. Kung, H. T. 1982. Why systolic architectures? Computer. January. pp. 37–46.
6. Whiteside, R. A., P. G. Hibbard & N. S. Ostlund. 1982. A systolic algorithm for Monte Carlo simulations. Proc. 3d. Intern. Conf. Distrib. Sys. IEEE Computer Society. pp. 800–804.
7. Whiteside, R. A., P. G. Hibbard & N. S. Ostlund. 1983. "Conventional" and systolic parallel algorithms for Monte Carlo simulations of molecular motions. ACM Trans. Comp. Sys. In press.
8. Verlet, I. 1967. Computer experiments on classical fluids. I. Thermodynamic

properties of Leonard-Jones molecules. Phys. Rev. **159:** 98–103.
9. WHITESIDE, R. A., N. S. OSTLUND & P. G. HIBBARD. 1984. A parallel diagonalization algorithm for a loop multiple processor system. IEEE Trans. Comp. **C-33:** 409–413.
10. WHITESIDE, R. A., P. G. HIBBARD & N. S. OSTLUND. 1984. Parallel algorithms for molecular dynamics simulations of molecular motion. In preparation.
11. FULLER, S. H., J. K. OUSTERHOUT, L. RASKIN, P. I. RUBINFELD, P. J. SINDHU & R. J. SWAN. 1978. Multi-microprocessors: An overview and working example. Proc. IEEE **66:** 216–228.

Index of Contributors

Andose, J. D., 124–139

Balaji, V. N., 140–161
Bing, D. H., 12–43
Bush, B. L., 124–139

Caporale, L. H., 12–43
Crippen, G. M., 1–11

Dixon, J. S., 140–161

Feldmann, R. J., 12–43
Fluder, E. M., 124–139
Furie, B., 12–43
Furie, B. C., 12–43

Greer, J., 44–63
Gund, P., 124–139

Hagler, A. T., 81–96
Hangauer, D. G., 124–139

Karplus, M., 107–123
Kowalczyk, P., 64–80

Mainhart, C., 12–43
Marshall, G. R., 162–169
McIntyre, E. F., 124–139
Miller, K. J., 64–80
Murdock, K. C., 140–161

Ostlund, N. S., 195–208

Potter, M., 12–43

Rein, F. H., 64–80
Rivier, J., 81–96

Salemme, F. R., 97–106
Scheraga, H. A., 170–194
Smith, D. H., 140–161
Smith, G. M., 124–139
Struthers, R. S., 81–96

Taylor, E. R., 64–80

Venkataraghavan, R., 140–161

Whiteside, R. A., 195–208